The Biogeochemical Cycling of Sulfur and Nitrogen in the Remote Atmosphere

NATO ASI Series

Advanced Science Institutes Series

A series presenting the results of activities sponsored by the NATO Science Committee, which aims at the dissemination of advanced scientific and technological knowledge, with a view to strengthening links between scientific communities.

The series is published by an international board of publishers in conjunction with the NATO Scientific Affairs Division

A	Life Sciences	Plenum Publishing Corporation
B	Physics	London and New York
C	Mathematical and Physical Sciences	D. Reidel Publishing Company Dordrecht, Boston and Lancaster
D	Behavioural and Social Sciences	Martinus Nijhoff Publishers
E	Engineering and Materials Sciences	The Hague, Boston and Lancaster
F	Computer and Systems Sciences	Springer-Verlag
G	Ecological Sciences	Berlin, Heidelberg, New York and Tokyo

Series C: Mathematical and Physical Sciences Vol. 159

The Biogeochemical Cycling of Sulfur and Nitrogen in the Remote Atmosphere

edited by

James N. Galloway
Department of Environmental Sciences, University of Virginia,
Charlottesville, Virginia, U.S.A.

Robert J. Charlson
Department of Civil Engineering, Environmental Engineering and Science Program,
University of Washington, Seattle, Washington, U.S.A.

Meinrat O. Andreae
Department of Oceanography, Florida State University,
Tallahassee, Florida, U.S.A.

and

Henning Rodhe
Department of Meteorology, Arrhenius Laboratory,
Stockholm University, Stockholm, Sweden

Technical editor:

Mary-Scott Marston

D. Reidel Publishing Company

Dordrecht / Boston / Lancaster / Tokyo

Published in cooperation with NATO Scientific Affairs Division

Proceedings of the NATO Advanced Research Workshop on
The Biogeochemical Cycling of Sulfur and Nitrogen in the Remote Atmosphere
St. Georges, Bermuda
8-12 October 1984

Library of Congress Cataloging in Publication Data

NATO Advanced Research Workshop on the Biogeochemical Cycling of Sulfur and Nitrogen
in the Remote Atmosphere (1984 : Bermuda Biological Station)
The biogeochemical cycling of sulfur and nitrogen in the remote atmosphere.

(NATO ASI series. Series C, Mathematical and physical sciences; vol. 159)
Proceedings of the NATO Advanced Research Workshop on the Biogeochemical Cycling
of Sulfur and Nitrogen in the Remote Atmosphere, held at the Bermuda Biological Station,
St. Georges, Bermuda, Oct. 8–12, 1984.
"Published in cooperation with NATO Scientific Affairs Division."
Includes indexes.
1. Sulphur cycle—Congresses. 2. Nitrogen cycle—Congresses. 3. Atmosphere—
Congresses. I. Galloway, James, 1944-   . II. North Atlantic Treaty
Organization. Scientific Affairs Division. III. Title. IV. Series: NATO ASI series.
Series C, Mathematical and physical sciences; vol. 159.
QH344.N38   1984   574.5'222   85-19905

ISBN-13: 978-94-010-8918-0   e-ISBN-13: 978-94-009-5476-2
DOI: 10.1007/978-94-009-5476-2

Published by D. Reidel Publishing Company
P.O. Box 17, 3300 AA Dordrecht, Holland

Sold and distributed in the U.S.A. and Canada
by Kluwer Academic Publishers,
190 Old Derby Street, Hingham, MA 02043, U.S.A.

In all other countries, sold and distributed
by Kluwer Academic Publishers Group,
P.O. Box 322, 3300 AH Dordrecht, Holland

D. Reidel Publishing Company is a member of the Kluwer Academic Publishers Group

Authors

Meinrat O. Andreae
Leonard A. Barrie
Bernard Bonsang
William L. Chameides
Robert J. Charlson
Robert B. Chatfield
Paul J. Crutzen
Robert J. Delmas
Sherry O. Farwell
James P. Friend
Ian E. Galbally
James N. Galloway
Lennart Granat

Gode Gravenhorst
Ivar S. A. Isaksen
Wolfgang Jaeschke
William C. Keene
Dieter Kley
Gene E. Likens
John M. Miller
Joseph M. Prospero
Henning Rodhe
Harold I. Schiff
F. Barry Smith
Elvira Tsani
Douglas M. Whelpdale

Observers

W. M. Ollison
Russell S. Drago (not pictured)

Bottom row (left to right): Tsani, Miller, Barrie, Schiff, Andreae,
     Charlson, Galloway, Rodhe
Second row: Keene, Friend, Ollison (Observer), Crutzen
Third row: Granat, Whelpdale, Marston, Likens, Jaeschke
Fourth row: Delmas, Bonsang, Kley, Prospero, Knap
Fifth row: Gravenhorst, Chatfield, Galbally, Smith, Isaksen

TABLE OF CONTENTS

LIST OF FIGURES

LIST OF TABLES

PREFACE

Viewed from space, the Earth appears as a globe without a beginning or an end. Encompassing the globe is the atmosphere with its three phases—gaseous, liquid, and solid—moving in directions influenced by sunlight, gravity, and rotation. The chemical compositions of these phases are determined by biogeochemical cycles.

Over the past hundred years, the processes governing the rates and reactions in the atmospheric biogeochemical cycles have typically been studied in regions where scientists lived. Hence, as time has gone by, the advances in our knowledge of atmospheric chemical cycles in remote areas have lagged substantially behind those for more populated areas. Not only are the data less abundant, they are also scattered. Therefore, we felt a workshop would be an excellent mechanism to assess the state-of-knowledge of the atmospheric cycles of sulfur and nitrogen in remote areas and to make recommendations for future research.

Thus, a NATO Advanced Research Workshop "The Biogeochemical Cycling of Sulfur and Nitrogen in the Remote Atmosphere" was held at the Bermuda Biological Station, St. Georges, Bermuda, from 8–12 October 1984. The workshop was attended by 24 international scientists known for their work in atmospheric cycling in remote areas. This volume contains the background papers and the discussions resulting from that workshop. The workshop was organized along the lines of the atmospheric cycle. There were working groups on emission, transport, transformation, and deposition. Towards the end of the week, an integration of the four groups was achieved by condensing the workshop into two groups, one on the S cycle and one on the N cycle. Since the scope of the workshop was on atmospheric S and N cycles in remote areas, we did not concern ourselves with the influence of man's activities on these cycles.

This book presents the groups' assessments of what was known and what needed to be known about S and N cycling in the remote atmosphere. The book should be of interest to several communities: (1) colleagues in atmospheric sciences and those concerned with atmospheric-biospheric interactions, (2) research managers requiring an overall assessment of the state-of-the-knowledge, and (3) individuals or agencies concerned with the scientific and political policy issues involving atmospheric cycling.

In addition to primary support from NATO, supplemental funding was provided by the Coordinating Research Council, British Petroleum International, the Bermuda Biological Station, the Government of Bermuda, the U. S. National Science Foundation, and the U. S. National Oceanic and Atmospheric Administration.

The program for the workshop was created by the Organizing Committee, which consisted of J. N. Galloway, Director (U.S.A.); R. J. Charlson,

Co-Director (U.S.A.); M. O. Andreae (U.S.A.), A. H. Knap (Bermuda), and
H. Rodhe (Sweden). As organizers of the workshop and editors of the
workshop report, we were fortunate to have worked with a wide variety of
competent individuals who helped make the workshop a pleasurable opportunity for scientific advancement. The Bermuda Biological Station was an
ideal setting because of its fine facilities and its outstanding staff.
Our special appreciation certainly goes to Harry Barnes and Jill Cad-
wallader for their happy help, patience, and resourcefulness, especially
at odd hours. Dr. Douglas Whelpdale, Atmospheric Environment Service,
Canada, provided the initial idea for the workshop; we are infinitely
grateful. Mary-Scott Marston served with great humor and brilliance, not
to mention patience, in her role as technical editor for the workshop and
the book. Brenda Morris's able hand at the word processor took our
scribbles and made them into legible reading material and Betsy Blizard's
graphics greatly enhanced our ideas. Also our thanks to William C. Keene
for his willing and constant support and assistance both in Bermuda and
Charlottesville. And last but not least, we cannot downplay the role
that wet deposition played in the success of the meeting--it rained the
entire time. Although this kept the atmosphere moist and writing paper a
bit damp, it also kept the participants off the beaches and in the
meeting rooms. Contrary to popular conception, the Organizing Committee
did not select the climatologically wettest week.

ABSTRACT

Before the meeting, invited participants were evenly divided into four working groups (Emission, Transport, Transformation, Deposition), each with its own chairman, and asked to review a background paper prepared by their individual chairman. These papers formed the basis for workshop discussions in which all members were expected to participate. At the meeting, all participants also contributed to two further working groups, one on sulfur cycling and one on nitrogen cycling.

The emission group concentrated on identifying and quantifying the natural sources of S and N species in remote marine and continental areas. Since estimates of emission rates by species for many areas of the earth were extrapolated from the few known point measurements, uncertainties were great. In some cases, it was not clear whether a particular region was a source or sink for certain gases.

The transport group characterized various "compartments" of the earth in terms of meteorological regimes to facilitate descriptions of chemical behavior. This group also proposed experimental means of investigating some unresolved cycling issues by identifying appropriate meteorological regimes in which to make measurements.

The transformation group examined the most likely transformation pathways for S and N species in gas, liquid, and aerosol phases and the interactions between the two cycles and among the various phases. Available field and laboratory data and chemical models for clean atmospheres were examined.

The deposition group concentrated their efforts on gathering and assessing all wet-deposition measurements for remote areas and found few high-quality measurements available. A method to measure dry deposition routinely had not yet been found. Locations and species for future measurements of wet and dry deposition to remote areas were suggested.

The two sessions on S and N cycling used the results from the four earlier groups to describe what was known about and what was needed to better understand each cycle. Using the regimes suggested by the transport group, available estimates of emission and deposition rates were compared.

A synthesis of state-of-the-art at the time of the workshop was produced and a series of thoughtful suggestions were agreed upon for the next steps necessary to improve our understanding of the biogeochemical cycling of S and N in remote areas.

INTRODUCTION

The atmosphere is a dynamic reservoir and the substances emitted to it
can be transported thousands of kilometers within the residence time of
most aerosols and soluble gases. Thus, the emissions in one region can
easily control the composition of the atmosphere and the atmospheric
deposition in another region. Because of the importance of the atmosphere and of its input and output fluxes to biogeochemical cycling, it
is incumbent upon us that we not only know the composition of the atmosphere and the magnitude of its fluxes but also the linkages between the
inputs, outputs, and reservoirs in between. Our knowledge of these
matters for the remote atmosphere has been limited by the paucity of data
on the composition and fluxes over scales of time and space long enough
to account for the large natural variability. This is unfortunate
because a knowledge of the biogeochemical cycling of S and N in remote
areas could provide a baseline for understanding the processes controlling the atmospheric composition in areas uninfluenced by man's
activities.

A knowledge of the atmospheric inputs and outputs is also required
for reasons beyond understanding atmospheric processes. Inputs and outputs come from and go to other reservoirs. Inputs are the results of
processes occurring in terrestrial and aquatic ecosystems. Measurements
of these inputs help establish the processes that control the nutrient
flow in these ecosystems. Atmospheric outputs impact terrestrial and
aquatic ecosystems. Measurements of these fluxes will enhance our understanding of the role that atmospheric deposition plays in the functioning
of the ecosystem.

The general objective of this workshop was to synthesize the available data on the chemical and physical linkages between the emission and
deposition of S and N species in remote areas. The chemical species
considered were S and N in their primary oxidation states. A motive for
understanding these linkages was to determine the magnitude of natural
fluxes and also the degree of alteration caused by man's activities. The
specific objectives of this workshop were:

- Objective 1: To estimate the rates of emission of S and N species in remote marine and continental areas and to determine
  the uncertainties of these estimates and the causes of those
  uncertainties.

- Objective 2: To estimate the influence on S and N species of
  long-range transport to remote areas.

J. N. Galloway et al. (eds.), The Biogeochemical Cycling of Sulfur and Nitrogen in the Remote Atmosphere, 1–2.
© *1985 by D. Reidel Publishing Company.*

- Objective 3: To determine the atmospheric processes that are important in controlling the transformation of these species in the atmosphere.

- Objective 4: To determine the best estimates of wet- and dry-deposition rates of S and N and their concentrations in wet deposition.

In summary, given the recent additions to the global data base on emission rates, atmospheric concentrations, and atmospheric deposition rates in remote areas, the time was appropriate to adopt an integrated approach to biogeochemical cycling in remote areas. This approach was process-oriented and concentrated on identifying the linkages between the emission to, the transport through, the transformation in, and the deposition from the remote atmosphere.

James N. Galloway
Robert J. Charlson
Meinrat O. Andreae
Henning Rodhe

PART I

THE EMISSION OF SULFUR AND NITROGEN TO THE REMOTE ATMOSPHERE

1. THE EMISSION OF SULFUR TO THE REMOTE ATMOSPHERE: BACKGROUND PAPER

M. O. Andreae
Department of Oceanography
Florida State University
Tallahassee, FL 32306
U.S.A.

1.1. INTRODUCTION

Both natural and anthropogenic sources contribute to the global atmospheric budget of sulfur. Anthropogenic sulfur emissions are dominated
by the release of SO_2 from fossil-fuel burning. Several authors have
recently reviewed these emissions and presented detailed source alloca-
tions (e.g., Möller 1984 and Cullis and Hirschler 1980). The estimates
from these papers fall into a relatively narrow range of about 70 Tg to
100 Tg S/yr.
     Various attempts to derive a global atmospheric budget for sulfur
have suggested that natural emissions of a magnitude comparable to man-
made ones are necessary to balance this budget (for a review, see Freney
et al. 1983). However, since the calculation of natural emissions from
the difference between estimates of anthropogenic emissions and total
deposition fluxes involves very large uncertainties, it cannot be expected to provide a meaningful estimate of the size of the natural emissions. This paper is based on the results of efforts to measure these
natural emissions to remote marine and continental environments.
     In this paper I present an overview of the processes that result in
the natural emissions of sulfur species and review the current best
estimates of the size of these fluxes. In some cases, e.g., the fluxes
resulting from biomass burning, the anthropogenic component results from
a modification of the type or size of a natural process. I emphasize the
natural component in these instances but I also discuss the cause and
magnitude of the anthropogenic perturbation.

1.2. NATURAL EMISSIONS OF SULFUR SPECIES

1.2.1. Aeolian Weathering of Sulfates in Arid Regions

Arid regions cover about $37 \cdot 10^6$ km^2, or about 10% of the earth's land
surface. Estimates of dust emissions from these regions vary between
200 Tg/yr and 3,000 Tg/yr (reviewed in Ryaboshapko 1983). Depending on
the assumed values for both the total dust mobilization and the sulfur
content of the dust, source estimates between 3 Tg and 30 Tg S/yr have

5

J. N. Galloway et al. (eds.), The Biogeochemical Cycling of Sulfur and Nitrogen in the Remote Atmosphere, 5–25.
© 1985 by D. Reidel Publishing Company.

been proposed. Substantial amounts of this dust can be transported over
more than 1,000 km and can be of great importance to regional sulfur
cycling (Ryaboshapko 1983). The spread of arid lands as a consequence of
land-use practices may substantially influence this flux.

1.2.2. Sea-salt Sulfate

The only significant direct source of particulate sulfur to the marine
atmosphere is from the production of sea-salt aerosol at the ocean surface. This aerosol is produced from seawater droplets that form when air
bubbles burst at the sea surface. Under very high wind conditions, droplets can also be formed when water is torn away from the crest of a wave.
Two types of droplets are produced by the bubble-bursting process: (1)
jet droplets from the breakup of a water jet rising from the bottom of
the collapsing bubble cavity and (2) film droplets from the bursting of
the thin film of water that separates the air in a bubble from the
atmosphere (Blanchard 1983). Partial evaporation of water from a seawater droplet produces brine drops or sea-salt particles, depending on
the relative humidity and the droplet diameter.
    The major ion composition of the sea-spray aerosol is generally
similar to that of seawater, but some evidence suggests that the sulfate
ion is enriched during the spray process. In model experiments, Garland
(1981) has observed an enrichment of sulfate of about 20% relative to
sodium but Duce et al. (1982) point out the possibility of artifacts in
Garland's experiments. Since secondary atmospheric processes can also
lead to a sulfate enrichment of sea-salt aerosol through the deposition
of sulfuric acid formed in the atmosphere (Andreae 1982), enrichment
during the spray-formation process cannot be easily deduced from field
studies of marine aerosols. Therefore, the question of whether at least
some sulfate enrichment in marine aerosols is caused by sea-surface
fractionation processes still remains open (Andreae 1982).
    To estimate the global source flux of sea-salt sulfate to the atmo-
sphere, we could start with an estimate of the flux of sea-salt aerosol
and predict the sea-salt-sulfate flux from the total sea-salt flux. This
would present some conceptual difficulty because the distribution of sea
salt in the marine atmosphere decreases steeply with increasing altitude
(Blanchard et al. 1984). Consequently, the flux through the lowest meter
of the atmosphere must be far in excess of 10^4 Tg/yr whereas the amount
of sea salt cycled through the upper troposphere should be several orders
of magnitude smaller (Blanchard 1983). Petrenchuk (1980) estimates a sea-
salt flux of approximately 1,300 Tg/yr through the lower troposphere (the
''subcloud layer''). By using this latter value plus a 20% enrichment of
sulfate during spray formation, we would obtain a flux of 40 Tg S/yr. On
the other hand, if we were to use Blanchard's value of 10^4 Tg/yr, the
predicted sulfur flux would be approximately 300 Tg S/yr. Therefore, one
can only state that the sulfate flux from the formation of sea-spray
aerosol is probably between 40 Tg and 300 Tg S/yr. Both estimates are
based on evaluations of sea salt found in wet- and dry-deposition
samples.
    In the past, rather limited data bases on deposition and rainwater
composition have been used to extrapolate to a global estimate, which has

resulted in large uncertainties. In recent years, we have learned much
more about the composition of marine rain and the dry deposition of sea
salt (McDonald et al. 1982, Galloway et al. 1982, Galloway et al. 1983,
Church et al. 1982). These newer data should considerably increase the
accuracy of the estimates of sea-salt flux.

   Most sea salt entering the atmosphere is redeposited to the ocean
surface. However, as much as 10% of the total flux is carried over
continents and deposited on land. Varhelyi and Gravenhorst (1983) have
estimated the flux of sea-salt sulfate using a compilation of sulfate
data from marine-aerosol and wet-deposition measurements. They give a
range of 120 Tg to 250 Tg S/yr for global sea-salt sulfate deposition,
which is consistent with the higher estimates for sea-salt flux from
Blanchard (1983). It must be emphasized that no actual emission measurements have yet been attempted. All available estimates of the flux of
sea-salt sulfate are based on aerosol concentrations and on deposition
data.

1.2.3. Volcanic Emissions of Sulfur Compounds

Volcanoes and geothermal areas emit a number of sulfur gases as well as
sulfate aerosol during both eruptive and noneruptive phases. Early
estimates of the size of the volcanic sulfur flux are quite low: 0.75 Tg
to 1.5 Tg S/yr (Kellogg et al. 1972 and Granat et al. 1976, respectively). These estimates are based on the gas content of the lava and
consider only SO_2 emissions; they also do not account for emissions
during noneruptive phases. Since 1972, sulfur-gas emissions have been
directly measured using remote-sensing techniques (correlation spectrom-
etry and aircraft data. In addition to SO_2, other sulfur species
(including H_2S, COS, and aerosol sulfate) have been found in volcanic
emissions.

   Based on a rather limited data set from 10 volcanoes and extrapolated for a global population of 578 active volcanoes, Lein (1983)
estimates a flux of 27 Tg S/yr from volcanoes (largely as SO_2) during
noneruptive periods. Lein's estimate of emissions during eruptions
(1.4 Tg/yr) is based on estimates of the volume of lava degassed during
eruptions and its volatile content. In a recent reassessment of the
volcanic contributions to the global sulfur cycle, Berresheim and
Jaeschke (1983) review the results of field measurements from volcanoes
throughout the world and conclude that SO_2 emissions during noneruptive
phases (8 Tg S/yr) are considerably larger than those during eruptive
periods (1 Tg S/yr). Hobbs et al. (1982) have reached similar conclusions about the emissions from Mt. St. Helens. Recent satellite measurements of SO_2 emissions during volcanic eruptions show that very large
amounts of sulfur can be emitted during a short period. For example,
during the eruption of 3 and 4 April 1982, El Chichon released about
1.6 Tg S in the form of SO_2 (Krueger 1983). As more data become available, an upward revision of the estimate of the eruptive sulfur flux may
become necessary.

   Very few data are available on the emissions of hydrogen sulfide
from volcanoes. Both Stith et al. (1978) and Berresheim and Jaeschke
(1983) estimate about 1 Tg S/yr. Again, most emissions they have studied

take place during noneruptive periods. However, their results are to be
expected on the basis of thermodynamic considerations since H_2S becomes
unstable in the presence of O_2 at elevated temperatures. During some
eruptions, however, H_2S appears to be dominating the emissions, e.g., at
Mt. St. Helens in 1980 (Hobbs et al. 1982) and during the 1982 eruption
of El Chichon (Kotra et al. 1983).

        Carbonyl sulfide (COS) and carbon disulfide (CS_2) have also been
observed in volcanic emissions. Cadle (1980) estimates the COS flux is
approximately 0.01 Tg S/yr. Based on their measurements at Mt. St.
Helens, Rasmussen et al. (1982a) estimate a similar global source
strength for CS_2.

        Several studies have reported the direct emission of sulfate aerosol
together with gaseous emissions. However, it is difficult, on the basis
of plume measurements, to discriminate between primary releases of sulfate from a volcano and the formation of sulfate by oxidation within a
plume. The ratio between the amounts emitted as sulfur gases and those
emitted as particulate sulfate is quite variable, even for a single volcano, but the gaseous emissions generally seem to dominate (Hobbs et al.
1982, Radke 1982, Kotra et al. 1983, Berresheim and Jaeschke 1983, and
references therein). Mostly from their work at Mt. Etna, Berresheim and
Jaeschke (1983) estimate a global particulate-sulfur flux from volcanoes
of as much as 3 Tg S/yr. This estimate is quite uncertain and may be
somewhat high. The estimates for SO_2, H_2S, and sulfate fluxes add up to
a total volcanigenic sulfur source to the atmosphere of about 12 Tg S/yr.

1.2.4. Production of Volatile Biogenic Sulfur Compounds

Sulfur is an essential element to biological organisms. Two biological
pathways lead from sulfate (the major sulfur source for almost all organisms) to reduced sulfur compounds: assimilatory and dissimilatory sulfate
reduction. Misunderstandings of the conceptual differences between these
two pathways have led to many misinterpretations and false assumptions in
the literature on the atmospheric sulfur cycle, such as the assumption
that H_2S is the major reduced-sulfur compound emitted from an ocean. In
assimilatory sulfate reduction, sulfate is taken up by the cell, acti-
vated to adenylyl sulfate, bound to a carrier molecule, and then reduced
stepwise to the -II oxidation state. The first major organic sulfur
metabolite is the amino acid cysteine, which is the starting compound
for the biosynthesis of the large variety of organosulfur compounds found
in biota. These sulfur compounds play essential biochemical roles as
components of proteins (enzymes) as well as in the methyl-group transfer
via S-adenosylmethionine and in structural materials. Consequently,
sulfur is found in all organisms at levels of a few tenths of a percent
(Bowen 1979).

        Volatile organosulfur compounds are produced from nonvolatile sul-
fur metabolites in biota either through the formation and release of
volatile sulfur metabolites in healthy, living cells or as a consequence
of decomposition processes. The most important product released from
live cells, especially by plants, is dimethylsulfide (DMS), which is
discussed in some detail in section 1.2.6.1. Decomposition of organic
matter by microbes leads to a variety of organosulfur compounds (Bremner

and Steele 1978, Kadota and Ishida 1972, Zinder et al. 1977). Free amino
acids, including those containing sulfur (cysteine, methionine, and homo-
cysteine), are released by the breakdown of proteinaceous materials. A
number of volatile compounds are produced during further decomposition,
many of which have strong odors (methylmercaptan and some higher mercap-
tans, dimethylsulfide and dimethylpolysulfide, carbon disulfide, and
several minor compounds). Hydrogen sulfide is also formed during the
decomposition of organosulfur compounds (rotting eggs!) and can contribute a significant fraction of the H_2S in anoxic environments (3% to 50%).

The major pathway to the formation of H_2S, however, is through
dissimilatory sulfate reduction. In contrast to the assimilatory pathway, the objective of which is the biosynthesis of sulfur compounds,
dissimilatory sulfate reduction is used to obtain thermodynamic energy in
an oxygen-depleted environment. The oxidation of organic matter by
available electron acceptors is the energetic basis for almost all life
processes. Thermodynamically, the most favorable electron acceptor is
molecular oxygen, and, if available, it will be used preferentially in
any ecosystem. However, if the supply of organic compounds exceeds that
of the oxygen, other electron acceptors, such as nitrate or sulfate, will
be used when the oxygen has been depleted. Therefore, dissimilatory
sulfate reduction is most commonly found in marine environments where
water circulation is limited (e.g., in stratified basins or in sedimen-
tary pore waters) but where sulfate is easily available because of its
relatively high concentration in seawater (28 mmol/kg).

Under favorable conditions, the rate of sulfate reduction to H_2S can
be quite high, on the order of hundreds of $mmol/(m^2 \cdot day)$. However,
this process only occurs when a mixing barrier prevents oxygen from
entering the system--the escape of H_2S from the system is, of course,
limited by the same barrier. Furthermore, in the presence of oxygen, H_2S
provides an excellent substrate for microbial oxidation from which certain bacteria can obtain a substantial amount of energy. Such microorganisms tend, therefore, to be present in high numbers at the oxic/
anoxic interface. They are very efficient in removing H_2S and can completely oxidize this compound in a layer only a fraction of a millimeter
thick. Consequently, the large amounts of H_2S that are produced in the
coastal and marine environment are not usually transferred to the atmosphere (Andreae 1984). Only a fraction of the H_2S can escape under
exceptional conditions in shallow water environments (e.g., through a
temperature or wind-driven turnover in estuaries, through the scouring of
mud in tidal channels, through the bubbling of gas from anoxic environments, etc.). H_2S emissions from the marine environment are, therefore,
limited to such nearshore environments as estuaries and salt marshes.

1.2.5. Biogenic Sulfur Emissions from the Continents

As recently as 1980, estimates of the magnitude of the biogenic emissions of sulfur gases were largely based on circumstantial reasoning
rather than on direct measurement (Cullis and Hirschler 1980). These
efforts resulted in estimates that vary between 30 Tg and 300 Tg S/yr for
total biogenic emissions and between 3 Tg and 110 Tg S/yr for emissions
from the land. In this paper, I focus on the results of direct measurements of emissions that have become available during the last few years.

In this section, I discuss emissions from inland soils, terrestrial plants, and freshwater and saline swamps.

1.2.5.1. **Emissions from Inland Soil and Plant Systems.** The only comprehensive study of sulfur gas emissions from inland soils is that of Adams et al. (1981). In addition to determining the fluxes of H_2S, COS, methylmercaptan, DMS, CS_2, and dimethyldisulfide (DMDS) from soils at 35 sites in the eastern and southeastern United States, Adams and his colleagues report several unidentified peaks. Since the flame photometric detector used in their work is not entirely sulfur-specific, these peaks may have been caused by large amounts of nonsulfur compounds.

The flux data show large variabilities between sites and between sampling periods at the same site (several orders of magnitude). For nonsaline soils they range from 0.002 g to 0.33 g $S/(m^2 \cdot yr)$ and for salt marshes the range is 0.03 g to 650 g $S/(m^2 \cdot yr)$. The average flux from inland soils is < 0.02 g $S/(m^2 \cdot yr)$ between the latitudes of 25° N to 47° N. Most flux from inland soils is in the form of H_2S (66%), the rest is accounted for by COS (13%), CS_2 (13%), DMS (7%), and a trace of DMDS (2%). Methylmercaptan (MeSH) appears not to be emitted from nonsaline soils. Because of the possibility of analytical error, the accuracy of the data for H_2S and MeSH are very uncertain (unpublished data available from Dr. S. O. Farwell, University of Idaho, Moscow, Idaho).

Most authors have determined the emissions of only H_2S. Delmas et al. (1980) have measured an average flux of 0.07 g $H_2S/(m^2 \cdot yr)$ from oxic lawn soils in France. Jaeschke et al. (1980) have found fluxes between 0.001 g and 1.1 g $H_2S/(m^2 \cdot yr)$ from marshland soils in the Ems River region of northern Germany. However, Delmas et al. (1980) report much higher fluxes from tropical soils on the Ivory Coast. They estimate that the mean annual flux from these soils ranges from 0.3 g to 0.9 g $H_2S/(m^2 \cdot yr)$. DMS fluxes from a variety of soils in Germany have been determined by Bürgermeister (data available from Dr. Bürgermeister, University of Frankfurt). Their range of observed DMS fluxes is 0.0004 g to 0.04 g $S/(m^2 \cdot yr)$, depending on soil type and temperature. A typical flux from mineral soils is about 0.0015 g $S(DMS)/(m^2 \cdot yr)$. These results compare reasonably well with those of Adams et al. (1981).

However, the extrapolation of these data to global fluxes is extremely problematic. If the average flux from inland soils in the eastern United States (0.021 g $S/[m^2 \cdot yr]$ from Adams et al. 1981) is extrapolated to obtain a worldwide value and 10% of the land area is subtracted to account for arid regions, one would obtain a flux of only 2.8 Tg S/yr. However, Adams and his colleagues use the pronounced latitudinal dependence in their data set to extrapolate to higher and lower latitudes and thereby predict a global flux from the land surface of 64 Tg S/yr (including coastal wetlands). The only independent data to support the extrapolation of Adams et al. are those of Delmas et al. (1980) from the Ivory Coast. Adams et al. also do not account for the much lower emissions that must be expected from arid and desert regions. Their estimate of the global biogenic sulfur flux from the land surface must be considered very preliminary and is probably much too high.

Since the soil-emission measurements discussed above have generally been conducted over vegetated soils, they include potential emissions by living plants. Plants may emit relatively large amounts of sulfur gases, especially in the tropics. Winner et al. (1981) have shown that H_2S is emitted from the leaves of living tomato plants. They extrapolate the flux from their tomato plants to the global plant biomass and predict a flux of ~ 50 Tg S/yr. Even though such an extrapolation is highly questionable, it does indicate the potential importance of plants as sources of atmospheric sulfur. However, much of the emitted H_2S may be recycled within tropical forest stands (Brinkmann and Santos 1974).

Rasmussen (1974) has shown that fresh oak and pine leaves also emit DMS. Freshwater algae produce DMS as well, but our work in estuaries shows that the DMS emissions from freshwaters are probably much less important than those from oceans (Froelich et al. 1985). These biogenic sulfur emissions in tropical forests may explain the relatively low rainwater pH found at sites in a tropical rain forest (pH 4.0 to 5.4 at San Carlos, Venezuela, [Galloway et al. 1982]). Most of this acidity, however, is caused by short-chain aliphatic acids, especially formic and acetic acids, and not by sulfuric acid.

As I discuss below, COS is formed by photochemically initiated reactions from organic sulfur compounds and molecular oxygen in seawater (Ferek and Andreae 1984). Plant tissues contain all the ingredients for the same reactions to occur. Because of the abundance of terrestrial plant tissue exposed to sunlight in comparison to the smaller amount of dissolved organic sulfur in an ocean's surface, living plants may represent a tremendous source of COS unless the photochemical reactions responsible for the production of COS are enzymatically inhibited in live tissue.

1.2.5.2. **Emissions from Coastal Wetlands.** Emissions of sulfur gases from saline marshes and intertidal areas vary widely between sites and between seasons and even between days (Aneja et al. 1982, Hansen et al. 1978, Adams et al. 1981, Jaeschke et al. 1980, Steudler and Peterson 1984, Ingvorsen and Joergensen 1982). The results from flux-chamber measurements range from not being detectable to nearly 2000 g $S/(m^2 \cdot yr)$ during peak periods. Typical emission rates for various sulfur species are presented in Table 1-1. Many of the reported differences are caused by diurnal and seasonal variations. The production of oxygen by photosynthesis during daytime leads to the rapid oxidation of H_2S in surface layers of intertidal sediments and marshes thereby preventing most of the H_2S produced in the anoxic zone from escaping into the atmosphere. At nighttime, the anoxic zone moves to the sediment surface and the emission of H_2S increases by several orders of magnitude (Hansen et al. 1978, Revsbech et al. 1983). COS is produced from organosulfur compounds by photochemical reactions (Ferek and Andreae 1984). The strong diurnal variability in COS emissions from salt marshes reported by Carroll et al. (1982) may be partly related to the photochemical dependence of the production rate and partly to soil-temperature and tidal cycles. Seasonal variations result from the effects of temperature on the rate of microbial reactions, especially sulfate reduction, and from seasonal changes in the community structure of marsh and tidal-flat ecosystems.

To extrapolate these values to global estimates may be hopeless. It may be more promising to obtain flux estimates using the micrometeorological method, where gas concentrations are measured at various altitudes

Table 1-1. Emissions of Volatile Sulfur Compounds (g S/[m^2 · yr]) from
            Coastal Wetlands.

Compounds	Various Saline Wetlands	Salt Marshes	Salt Marshes
H$_2$S	0.6	0.55	2.4
COS	0.09	---	0.34
MeSH	0.09	≤1.9	---
DMS	0.22	0.66	1.5
CS$_2$	0.12	≤0.2	0.2
DMDS	0.01	≤0.81	0.35
Total S	1.1	<11	4.8

Sources: For saline wetlands, Adams et al. 1981; for salt marshes in the
         middle column, Aneja et al. 1982; for salt marshes in the right-hand
         column, Steudler and Peterson 1984.

above the source while accounting for wind speed and direction. The flux
could then be estimated by applying turbulent–diffusion fusion theory to
the concentration–profile data. This approach would eliminate the artifacts created by installing the flux chamber and would integrate the flux
over a larger area. Goldberg et al. (1981) have measured the differences
in atmospheric H$_2$S concentrations upwind and downwind from a coastal salt
marsh at Wallops Island, Virginia, and they estimate that H$_2$S emission
rates are near 0.4 g S/(m^2 · yr) in December and 6 g S/(m^2 · yr) in July.
     It is difficult to assign an average value to the flux of volatile
sulfur compounds from coastal wetlands. However, even if we were to use
a relatively high estimate of 10 g S/(m^2 · yr) for the average annual
flux and combine it with the total global coastal wetland area of
380,000 km, we would obtain a global flux of ~ 4 Tg S/yr, which is
only a small percentage of the total biogenic emissions.

1.2.5.3. Emissions from Biomass Burning. Biomass burning is both a
natural and an anthropogenic source of sulfur. Most biomass is burned
because of agricultural practices and the use of wood as fuel (~ 5000 Tg
dry matter burned per year [dm/yr] and ~ 1000 Tg dm/yr, respectively);
wildfires contribute little to the total amount of biomass burned
(~ 250 Tg dm/yr, Seiler and Crutzen 1980). Since most of the burning
occurs in relatively undeveloped regions, especially in the tropics, I
have included a short discussion on its importance as a sulfur source.
     The average sulfur content of land plants is ~ 0.2% (Bowen 1979),
which corresponds to a potential sulfur flux of ~ 14 Tg S/yr at a burning
rate of 6800 Tg dm/yr. This estimate may be low since biomass burning
primarily affects the leafy parts of plants, which tend to have higher

sulfur concentrations than other parts. Not all the sulfur in the com-
busted biomass enters the atmosphere since a significant fraction is
retained in the ash (35%-67% in the case of the savannah fires described
by Delmas in 1982). Assuming an average volatilization efficiency of
50%, an estimate of 7 Tg S/yr was obtained for the total flux.

The chemical forms in which sulfur is emitted from fires and their
relative proportions are still quite uncertain. It has been assumed that
SO_2 was the major product, but H_2S, COS, and sulfate aerosol have also
been observed and may be important. The global emission rate for COS
from biomass burning has been estimated to be 0.11 Tg S/yr. Although
this is only a minor fraction of the sulfur flux from fires, it is a
significant component in the atmospheric cycle of COS (Crutzen et al.
1979). Sulfur gases and aerosols have not yet been measured together in
biomass-fire emissions. The results from one experiment where SO_2 and
particulate sulfur have been measured in emissions from a wood-burning
stove suggest that SO_2 may be dominant (Cooper 1980).

1.2.6. Emission of Volatile Sulfur Species from the Oceans

Oceans cover 71% of the earth's surface (61% in the Northern Hemisphere,
81% in the Southern). Therefore, even a relatively small flux of a
compound per unit area of ocean may translate into a significant global
flux. Volatile substances are transferred across the air/sea interface by
a combination of molecular and turbulent diffusion processes, which are
still poorly understood and for which no entirely satisfactory physical
or mathematical models are available. For a discussion of the state of
the art in this field, see the review by Liss (1983).

Fundamentally, the transfer flux can be expressed as the product of
a gradient (ΔC) and a proportionality constant (exchange coefficient or
transfer velocity, K),

$$F = K \cdot \Delta C. \tag{1-1}$$

The problem is in assigning values to K and ΔC. The gradient ΔC is best
represented by the difference in chemical potential between liquid and
gas phases. Since this thermodynamic function is usually not easily
obtainable, it is normally replaced by concentrations that are brought to
common units using Henry's Law constant, H,

$$F = K (C_g/H - C_L), \tag{1-2}$$

where C_g is the concentration in the gas and C_L, in the liquid phase.
The value of ΔC is especially difficult to determine for reactive gases
(e.g., SO_2) when the reactions in the liquid phase (in this case formation of bisulfite ion) are fast relative to the diffusion process. Because gases that are photochemically produced or destroyed may also show
serious gradients near the surface, the concentration gradient in measurements from water samples collected within a meter or more of the
surface is not representative.

The transfer velocity, K, for radon has been experimentally determined in field studies and in wind tunnels for CO_2 and, to a limited

extent, some other gases in wind tunnels. Bingemer (1984) has made the
only measurements of any sulfur gases in a chamber system at zero wind
speed using the DMS naturally present in seawater. For DMS concentrations of 40 ng to 900 ng S(DMS)/L, he found fluxes that compare reasonably well with the predictions from theory. For other gases, however, we
have to rely only on extrapolations based on relatively poorly developed
theories. The uncertainties involved in estimating transfer fluxes by
this model introduce a potential error of as high as a factor of two into
the oceanic source estimates of most compounds.

      Alternative methods to determine the flux (e.g., the eddy correla-
tion or gradient technique) have faced large experimental difficulties.
No rapid-response sensor that will make the eddy-correlation technique
possible is available for any sulfur gas in the remote atmosphere. Nguyen et al. (1984) have used the gradient method onboard ship by taking
samples at different levels above the waterline. Although the results
are comparable to the predictions from gas-transfer calculations, they
may contain substantial errors because of the influence of the ship on
airflow characteristics. Since a realistic wave climate is difficult to
simulate inside a flux chamber, sulfur-gas fluxes across the air/sea
interface have not been measured by the chamber technique. Therefore, in
the following discussion of the fluxes of individual sulfur species, I
have relied on predictions from the transfer model, which are based on
gas concentrations measured in seawater and boundary-layer air.

1.2.6.1. <u>Dimethylsulfide</u>. In open ocean waters, dimethylsulfide (DMS)
is the predominant volatile sulfur compound (Barnard et al. 1982, Andreae
et al. 1983, Andreae and Barnard 1984, Nguyen et al. 1984, Cline and
Bates 1983, Bingemer 1984). Although other compounds, especially CS_2 and
methylmercaptan, may also be present in coastal waters (Turner and Liss
1985), their limited distribution and relatively low concentrations allow
for only minor contributions to the global atmospheric sulfur budget.
DMS is produced in the surface ocean by phytoplankton from an intracellu-
lar biochemical precursor, dimethylsulfonium propionate (DMSP--also
called dimethyl propiothetin, DMPT). This compound has an osmoregulatory
function in some algal species (Vairavamurthy et al. 1985) and is enzymatically cleaved into DMS and acrylate. We do not yet know if DMS
itself has any physiological function in algae.

      The analysis of large data sets and the vertical distribution of DMS
in the marine water column clearly show that the DMS concentration in
seawater is related to the abundance of phytoplankton. Nevertheless, a
direct correlation between total plankton abundance and DMS concentration within a given region has not yet been established because of the
substantial differences in the DMS output rate of different plankton
species (Andreae et al. 1983, Barnard et al. 1984). In some cases, a
single phytoplankton species can be responsible for most of the DMS in a
given oceanic region, e.g., <u>Phaeocystis poucheti</u> in the Bering Sea shelf
region (Barnard et al. 1984). Our data also show that DMS concentrations
in low-productivity regions of oceans, especially the subtropical gyres,
are substantially higher than has been previously expected based on the
traditionally accepted primary productivity rates for these areas. This
may again be a species-related effect, or it may reflect the possibility

that the productivity in these regions is underestimated, as Shulenberger and Reid (1981) and Jenkins (1982) suggest. When emissions of DMS from different marine biogeographic regions are compared, emissions from large oceanic regions of relatively low productivity appear to be as important to the global flux as those from localized ''hot-spots'' of high productivity, e.g., upwelling areas (Table 1-2). The most important gap in our knowledge of the distribution of DMS in the world's oceans is in the circum-Antarctic region. The reported high rates of productivity in regions of the southern oceans and high ventilation rates caused by strong winds may result in substantial DMS fluxes.

Based on a large data set from a wide variety of oceanic biogeographical zones, Andreae and Raemdonck (1983) have calculated an average DMS concentration for each zone. Using data on the areal extent of these zones, we have determined a weighted global average for the DMS concentration in surface seawater of ~ 100 ng S(DMS)/L (Table 1-2), from which we predict an average DMS flux of 0.11 g S/(m^2 · yr). This translates to a global sea-to-air flux of ~ 40 Tg S/yr. This estimate is subject to an uncertainty of about ± 20 Tg, most of which derives from the uncertainties involved in using the ''stagnant-film'' model to estimate air-sea transfer. Using the gradient method mentioned above, Bingemer (1984) has estimated a flux of 0.26 g S/(m^2 · yr) for a set of 10 measurements in a relatively productive region of the South Atlantic. This flux compares well with the fluxes predicted on the basis of his seawater concentration data from the transfer-velocity model. Other researchers have recently published comparable flux estimates based on geographically limited concentration data sets (Nguyen et al. 1984, Bingemer 1984, Cline and Bates 1983).

To verify these estimates, it is important to find independent checks on the flux of DMS from the oceans. We have recently obtained a large set of measurements on DMS from a number of sites in remote marine atmospheres (Cape Grim, Tasmania, the equatorial Pacific, the tropical Atlantic east of the Bahamas, and the temperate North Atlantic). When no strong pollutant influence is present in the air mass, DMS concentrations of 100-300 pptv are consistently found with clear diurnal variations characterized by an early-morning maximum and a late-afternoon minimum (Andreae and Raemdonck 1983, Andreae et al. 1985). These concentrations are consistent with the measured reaction rate of DMS with OH and a sea-to-air flux of the same magnitude as estimated above (Graedel 1979, Chatfield and Crutzen 1984). Lower atmospheric DMS concentrations obtained earlier (e.g., Barnard et al. 1982) are to a large part caused by an oxidant interference in the sampling device. In the presence of atmospheric pollutants, the DMS levels found in the atmosphere are usually considerably lower than those predicted by the photochemical model, possibly because of a nighttime reaction with the nitrate radical NO_3 (Winer et al. 1984). Atmospheric concentrations of DMS of a few hundred pptv are in turn consistent both with the amount of excess sulfate found in remote marine aerosol (Bonsang et al. 1980, Andreae 1982, Maenhaut et al. 1983) and with the levels of SO_2 found in the remote marine boundary layer (Maroulis et al. 1980). Estimates of the flux of natural excess sulfate in the marine atmosphere also require a source of reduced sulfur gases on the order of 0.04-0.08 g S/(m^2 · yr) or about half our estimated

Table 1-2. The DMS Concentrations and Sea-to-air Flux from the Major
    Ecological Regions of the Earth's Oceans.

Ocean Region	Average DMS (ng S/l)	Standard Deviation	Area (10^6 km^2)	Flux* (10^{12} g S/yr)
Oligotrophic areas				
Tropical North Atlantic	98.1	44.7		
Gulf of Mexico	52.3	9.8		
Gulf Stream, Sargasso Sea	80.2	22.8		
Brazil Current	46.0	15.3		
Tropical North Pacific	59.8	10.8		
Total	67.1		148.3	10 ± 5
Transitional areas				
North Atlantic (temperate)	66.8	28.3	82.8	6 ± 3
Upwelling areas				
Equatorial Zone				
Atlantic	70.1	23.3		
Pacific	120.4	46.4		
Peru Shelf	230.9	235		
Frontal Areas				
Bering Sea	150.1	152		
Ushant Front	156.9	38.0		
Rio de la Plata Estuary	571.0	237		
Total	177.1		86.5	16 ± 8
Coastal and Shelf Zone				
North Sea, English Channel	54.9	28.2		
South America (eastern coast)	159.4	12.5		
Ecuador shelf	175.4	123		
Total	136.6		49.4	7 ± 3.5
World Ocean Totals	102.4		367.1	40 ± 20

*Uncertainty in fluxes largely due to the method of flux estimation
from seawater concentrations.

DMS input flux (Rodhe and Isaksen 1980, Kritz 1982). Considering that
these estimates involve uncertainties and that DMS has sinks other than
excess sulfate (e.g., methane sulfonic acid and dimethyl sulfoxide),
these independent flux estimates agree well with the DMS flux estimated
above (see Chapter 10).

1.2.6.2. Carbonyl Sulfide. Concentrations of one to a few ng $S(COS)/L$ are found in surface seawater (Rasmussen et al. 1982b, Ferek and Andreae 1983 and 1984, Turner and Liss 1985). Since observed concentrations are almost always higher than the equilibrium concentration relative to the overlying atmosphere, a net sea-to-air flux exists essentially across the entire ocean surface. An undersaturation of COS in the ocean's surface has only been found in regions of low biological productivity, especially at night and during high wind conditions (Andreae and Ferek, unpublished data available from the author).

Carbonyl sulfide is produced from dissolved organic sulfur compounds in seawater by photochemical reactions (Ferek and Andreae 1984), resulting in strong diurnal variations in the concentration of COS in surface seawater. These reactions require dissolved molecular oxygen as well as light and organic sulfur. Although exact quantitative relationships have not yet been determined, we have found that the concentration of COS is roughly proportional to the integrated solar intensity and to the concentration of dissolved organic matter and inversely related to wind intensity. Because little or no COS is directly produced or consumed through any biological process in seawater, ventilation of the water column must be its major loss process. Johnson (1981) speculates that the ocean may be a sink for atmospheric COS because of its hydrolysis at the slightly alkaline pH of seawater. However, this suggestion is not supported by the recent measurements of the COS supersaturation ratio across the air/sea interface.

To estimate a realistic flux of COS from the oceans to the atmosphere, supersaturation must be integrated over the diurnal cycle and over the different light fluxes, the dissolved organic sulfur concentrations, and the specific wind conditions found in different ocean regions. Current estimates do not satisfy these requirements and can, therefore, only be considered rough guesses. Using samples collected without considering the diurnal cycle and only separated into coastal and open-ocean values, Rasmussen et al. (1982b) estimate a global COS flux from the oceans of ± 0.3 Tg S/yr. Based largely on coastal data and interpolation from an observed relationship of DMS and COS, Ferek and Andreae (1983) estimate a global flux of ± 0.5 Tg S/yr. Recent work by my group at Florida State University has considerably expanded the data base used in 1983, resulting in an estimate of 0.4 ± 0.2 Tg S/yr for the oceanic source of COS.

1.2.6.3. Carbon Disulfide. CS_2 in seawater was originally found by Lovelock (1974) who measured an average concentration of 0.44 ng $S(CS_2)/L$ in 35 samples taken from the open Atlantic. Inshore values were about an order of magnitude higher. Turner and Liss (1985) have also reported on CS_2 found in coastal waters off England. However, because of a chromatographic separation problem between DMS and CS_2, Turner and Liss do not present any quantitative information. I have attempted to establish a value for CS_2 in open-ocean seawater but have not yet found it at a detection limit of 0.8 ng $S(CS_2)/L$, which is consistent with Lovelock's results. Assuming an average concentration of 0.6 ng $S(CS_2)/L$ in surface seawater, I have estimated a flux on the order of 0.4 Tg S/yr in the form of CS_2 from the surfaces of the world's oceans, or about 1% of the DMS flux.

1.2.6.4. Methylmercaptan and Dimethyldisulfide. We have observed peaks for methylmercaptan and its oxidation product, dimethyldisulfide, in many seawater samples from coastal regions (unpublished data available from the author). Because of calibration problems, we have not yet been able to report any quantitative data but the concentrations of these compounds are always much smaller than those of DMS. We have not yet detected them in the open ocean. Although methylmercaptan and dimethyldisulfide probably do not make a major contribution to the global sulfur cycle, some quantitative data should be obtained on their concentrations in surface seawater.

1.2.6.5. Hydrogen Sulfide. In models of the atmospheric sulfur cycle that have been reported up to the last few years, it has always been assumed, without experimental justification, that H_2S has been responsible for essentially all the flux of reduced sulfur from the oceans to the atmosphere. Now, still without any data, a complete reversal has occurred and H_2S is often ignored altogether. The latter assumption is usually based on the relatively rapid oxidation of H_2S in seawater. Half-lives for H_2S in seawater on the order of a few hours have been reported by Almgren and Hagström (1974) although other researchers have found values as high as 50 hrs (Chen and Morris 1972). H_2S has been found in the marine atmosphere at levels from a few pptv to a few tens of pptv (Slatt et al. 1978, Delmas and Servant 1982, Herrmann and Jaeschke 1984). These papers do not clarify how much of their values for H_2S could be caused by the interference from DMS. Using these values and a photochemical model, Graedel (1979) states that a source of 15 Tg $S(H_2S)$/yr would be required to support this concentration. However, assuming a diurnally averaged OH concentration of $2 \cdot 10^6$ molec/cm^3, a reaction rate constant of $5 \cdot 10^{-12}$ cm^3/(molec $\cdot$ sec) (Cox and Sheppard 1980), and an average concentration of H_2S in the remote marine troposphere of 10 ng $S(H_2S)$/m^3 with a scale height of 2.5 km, we would obtain a flux of only about 2.5 Tg S/yr. It is not clear, however, whether the ocean's surface is the source of this H_2S or whether other processes are responsible for its presence. McElroy et al. (1980) speculate that reactions of COS and CS_2 with OH radical may produce the necessary amounts of H_2S but this suggestion has not yet been experimentally verified. (These reactions are discussed in Part II.) The concentration of H_2S in surface seawater and in the overlying atmosphere must be determined directly so that the magnitude and direction of the exchange flux (since the sea surface may well even be a sink for H_2S!) can be estimated.

1.2.7. Summary of Natural Sulfur Emissions

The fluxes of the sulfur compounds SO_2, H_2S, COS, DMS, CS_2, and sulfate are summarized in Table 1-3. Large uncertainties obviously exist about both the chemical speciation and the magnitude of these fluxes. Although the estimated sea-salt aerosol flux introduces a very large uncertainty into the total estimate, even after removing this component, the range of estimates for the gaseous flux is still 65-125 Tg S/yr. The most problematic estimates are for the possible emissions from live plants and those from inland soils. The estimates for inland soils suffer because

Table 1-3.   Estimates of Natural Sulfur Emissions (Tg S/yr).

	SO_2	H_2S	COS	DMS	CS_2	Sulfate	Other	Total
Sea spray						40–300		40–300
Dust						3–30		3–30
Total Particulates						40–330		40–330
Volcanoes	8	1	0.01		0.01	(<3)	?	12
Soils and								
plants	–	3–41	0.2–0.6 (8?)	0.2–4	0.6–0.8 (8?)	–	1	5–47
Coastal								
wetlands	–	0.9	0.13	0.6	0.07	–	0.13	1.8
Biomass								
burning*	7	?	0.11	–	?	?	?	7
Oceans (gases)	–	0–15	0.4	40	0.4	–		40–56
Total gases	15	5–58	0.8–1.2	40–45	1.1–1.3	(≤3)	1	65–125

*All gaseous emissions (other than COS) assumed to be SO_2.

the only representative data set comes from a relatively narrow latitudinal band with a strong latitude dependency, which suggests that most of the flux is taking place outside of the investigated region. Depending on how the fluxes are extrapolated for the tropics, an order-of-magnitude difference can result. If we use the average ratios of COS and CS_2 to total sulfur emissions as given by Adams et al. (1981) and apply these ratios to the estimate of total sulfur emitted from inland soils, we get the very high estimates of about 8 Tg S/yr for each compound indicated in parentheses in Table 1-3. These fluxes are much higher than can be accounted for in the global cycles of COS and CS_2. Obviously, research to investigate biogenic sulfur emissions from tropical terrestrial systems should have a very high priority. Direct evidence is also urgently needed to verify the potential sea-to-air flux of H_2S. The limited research on emissions from live plants, possibly a major S flux, needs to be expanded.

Upper limits on these fluxes might be established by examining the total rates of sulfate reduction by plants and algae. These emissions would have to be a ''reasonable'' fraction of the total amount of sulfur assimilated into biota. Table 1-4 shows an overview of the rates of sulfate reduction by different processes. The major reduction processes are (1) the reduction of seawater sulfate by Fe(II) in basalts as a result of ocean-floor hydrothermal circulation (This process is entirely decoupled from the atmosphere and can be ignored for our purposes.); (2) the bacterial, dissimilatory reduction of sulfate to hydrogen sulfide during anaerobic respiration, which takes place mainly in marine sediments and, for reasons, discussed above, little of the sulfide produced can escape

Table 1-4.   Sulfate Reduction Rates by Major Biogeochemical Processes.

Process	Rate (Tg S/yr)	Reference
Reduction by Fe(II) in submarine hydrothermal systems (upper limits)	≤120	Von Damm 1983
Bacterial, dissimilatory sulfate reduction		
Coastal zone	70	
Shelf sediments	190	
Slope sediments	280	
Total	400–600	Ivanov & Freney 1983
Assimilatory sulfate reduction		
Land plants	100–180	
Marine algae	300–600	
Total	400–800	Ehrlich et al. 1977

to the atmosphere; and (3) the assimilatory reduction of sulfate by plants and algae through which plants and algae incorporate divalent sulfur into organic compounds. This latter process appears to be the largest reduction flux and to provide most of the reduced sulfur that can enter the atmosphere in the form of biogenic volatiles.

Clearly, the amount of DMS entering the atmosphere from the oceans (~ 40 ± 20 Tg S/yr) is a reasonable fraction of the amount of sulfur fixed by marine algae (~ 300–600 Tg S/yr, Table 1-4). On the other hand, the sum of the high estimates of terrestrial emissions (56 Tg S/yr—49 Tg S/yr from soils and living plants and 7 Tg S/yr from biomass burning) appears to be much too high relative to the terrestrial sulfate reduction rate of ~ 100–180 Tg S/yr. A volatilization efficiency of about 20% may be more reasonable; this would lead to a prediction of 20–40 Tg S/yr for continental emissions. Therefore, ~ 70 Tg S/yr would be a plausible estimate for total biogenic emissions and, by using 12 Tg S/yr for volcanic emissions, one would obtain a natural global sulfur flux of about 80 Tg S/yr, or about the same as the anthropogenic SO_2 flux.

1.3.   REFERENCES

Adams, D. F., S. O. Farwell, E. Robinson, M. R. Pack, and W. L. Bamesberger. 1981. Biogenic sulfur source strengths. Environ. Sci. Technol. 15:1493–1498.
Almgren, T., and I. Hagström. 1974. The oxidation rate of sulphide in sea water. Water Res. 8:395–400.
Andreae, M. O. 1982. Marine aerosol chemistry at Cape Grim, Tasmania and Townsville, Queensland. J. Geophys. Res. 87:8875–8885.

Andreae, M. O. 1984. Atmospheric effects of microbial mats. In Micro-
    bial Mats: Stromatolites (Y. Cohen, R. W. Castenholz and H. O. Hal-
    vorson, eds.) New York:Alan R. Liss, 455-466.
Andreae, M. O., and W. R. Barnard. 1984. The marine chemistry of di-
    methylsulfide. Marine Chem. 14:267-279.
Andreae, M. O., and H. Raemdonck. 1983. Dimethylsulfide in the surface
    ocean and the marine atmosphere: A global view. Science 221:744-747.
Andreae, M. O., W. R. Barnard, and J. M. Ammons. 1983. The biological
    production of dimethylsulfide in the ocean and its role in the global
    atmospheric sulfur budget. Ecol. Bull. (Stockholm) 35:167-177.
Andreae, M. O., R. J. Ferek, F. Bermond, K. P. Byrd, R. T. Engstrom, S.
    Hardin, P. D. Houmere, F. LeMarreck, R. B. Chatfield, H. Raemdonck.
    1985. Dimethylsulfide in the marine atmosphere. J.Geophys. Res. (in
    press).
Aneja, V. P., A. P. Aneja, and D. F. Adams. 1982. Biogenic sulfur com-
    pounds and the global sulfur cycle. JAPCA 32:803-807.
Barnard, W. R., W. E. Watkins, H. Bingemer, and H. -W. Georgii. 1982.
    The flux of dimethylsulfide from the oceans to the atmosphere. J.
    Geophys. Res. 87:8787-8793.
Barnard, W. R., M. O. Andreae, and R. L. Iverson. 1984. Dimethylsulfide
    and Phaeocystis poucheti in the southeastern Bering Sea. Cont.
    Shelf Res. 3:103-113.
Berresheim, H., and W. Jaeschke. 1983. The contribution of volcanoes to
    the global atmospheric sulfur budget. J. Geophys. Res. 88:3732-3740.
Bingemer, H. G. 1984. Dimethylsulfid in Ozean und mariner Atmosphäre-
    Experimentelle Untersuchung einer natürlichen Schwefelquelle für die
    Atmosphäre. Ph.D. Diss., J. W. Goethe University, Frankfurt am Main.
Blanchard, D. C. 1983. The production, distribution, and bacterial enrich-
    ment of the sea-salt aerosol. In Air-Sea Exchange of Gases and Par-
    ticles (P.S. Liss and W.G.N. Slinn, eds.) Boston: Reidel, 407-454.
Blanchard, D. C., A. H. Woodcock, and R. J. Cipriano. 1984. The vertical
    distribution of the concentration of sea salt in the marine atmo-
    sphere near Hawaii. Tellus 36B:118-125.
Bonsang, B., B. C. Nguyen, A. Gaudry, and G. Lambert. 1980. Sulfate en-
    richment in marine aerosols owing to biogenic gaseous sulfur
    compounds. J. Geophys. Res. 85:7410-7416.
Bowen, H. J. M. 1979. Environmental Chemistry of the Elements. New York:
    Academic Press.
Bremner, J. M., and C. G. Steele. 1978. Role of microorganisms in the
    atmospheric sulfur cycle. In Vol. 2: Advances in Microbial Ecology
    (M. Alexander, ed.) New York:Plenum, 155-201.
Brinkmann, W. L. F. and V. de M. Santos. 1974. The emission of biogenic
    hydrogen sulphide from Amazonian floodplain lakes. Tellus
    26:262-267.
Cadle, R. D. 1980. A comparison of volcanic with other fluxes of atmos-
    pheric trace gas constituents. Rev. Geophys. Space Phys. 18:746-752.
Carroll, M. A., L. E. Heidt, R. J. Cicerone, and R. G. Prinn. 1982.
    Carbonyl sulfide fluxes from a salt-water marsh. EOS 63:893.
Chatfield, R. B., and P. J. Crutzen. 1984. Sulfur dioxide in remote oce-
    anic air: Cloud transport of reactive precursors. J. Geophys. Res.
    89:7111-7132.

Chen, K. Y., and J. C. Morris. 1972. Kinetics of oxidation of aqueous
    sulfide by O_2. Environ. Sci. Technol. 6:529-537.
Church, T. M., J. N. Galloway, T. D. Jickells, and A. H. Knap. 1982. The
    chemistry of western Atlantic precipitation at the mid-Atlantic coast
    and on Bermuda. J. Geophys. Res. 87:11,013-11,018.
Cline, J. D., and T. S. Bates. 1983. Dimethyl sulfide in the equatorial
    Pacific Ocean: A natural source of sulfur to the atmosphere. Geo-
    phys. Res. Lett. 10:949-952.
Cooper, J. A. 1980. Environmental impact of residential wood combustion
    emissions and its implications. JAPCA 30:855-861.
Cox, R. A., and D. Sheppard. 1980. Reactions of OH with gaseous sulphur
    compounds. Nature 284:330.
Crutzen, P. J., L. E. Heidt, J. P. Krasnec, W. H. Pollock, and W. Seiler.
    1979. Biomass burning as a source of atmospheric gases CO, H_2, N_2O,
    CH_3Cl and COS. Nature 282:253-256.
Cullis, C. F., and M. M. Hirschler. 1980. Atmospheric sulfur: natural and
    man-made sources. Atmos. Environ. 14:1263-1278.
Delmas, R. 1982. On the emission of carbon, nitrogen and sulfur in the
    atmosphere during bushfires in intertropical savannah zones. Geo-
    phys. Res. Lett. 9:761-764.
Delmas, R., and J. Servant. 1982. The origins of sulfur compounds in the
    atmosphere of a zone of high productivity (Gulf of Guinea). J. Geo-
    phys. Res. 87:11,019-11,026.
Delmas, R., J. Baudet, J. Servant, and Y. Baziard. 1980. Emissions and
    concentrations of hydrogen sulfide in the air of the tropical forest
    of the Ivory Coast and of temperate regions in France. J. Geophys.
    Res. 85:4468-4474.
Duce, R. A., F. MacIntyre, and B. Bonsang. 1982. Enrichment of sulfate in
    maritime aerosols. Atmos. Environ. 16:2025-2034.
Ehrlich, P. R., A. H. Ehrlich, and J. P. Holdren. 1977. Ecoscience: Popu-
    lation, Resources, Environment. San Francisco:Freeman, 1051 p.
Ferek, R. J., and M. O. Andreae. 1983. The supersaturation of carbonyl
    sulfide in surface waters of the Pacific Ocean off Peru. Geophys.
    Res. Lett. 10:393-396.
Ferek, R. J., and M. O. Andreae. 1984. Photochemical production of car-
    bonyl sulfide in marine surface waters. Nature 307:148-150.
Freney, J. R., M. V. Ivanov, and H. Rodhe. 1983. The sulfur cycle. In
    SCOPE 24: The Major Biogeochemical Cycles and Their Interactions
    (B. Bolin and R. B. Cook, eds.), New York: Wiley, 56-61.
Froelich, P. N., L. W. Kaul, J. T. Byrd, M. O. Andreae, and K. K. Roe.
    1985. Arsenic, barium, germanium, tin, dimethylsulfide, and nutrient
    biogeochemistry in Charlotte Harbor, Florida, a phosphorus-enriched
    estuary. Estuarine, Coastal and Shelf Sci., 20:(in press).
Galloway, J. N., G. E. Likens, W. C. Keene, and J. M. Miller. 1982. The
    composition of precipitation in remote areas of the world. J. Geo-
    phys. Res. 87:8771-8786.
Galloway, J. N., A. H. Knap, and T. M. Church. 1983. The composition of
    western Atlantic precipitation using shipboard collectors. J. Geo-
    phys. Res. 88:10,859-10,864.
Garland, J. A. 1981. Enrichment of sulphate in maritime aerosols. Atmos.
    Environ. 15:787-791.

Goldberg, A. B., P. J. Maroulis, L. A. Wilner, and A. R. Bandy. 1981. Study of H_2S emissions from a salt water marsh. Atmos. Environ. 15:11-18.

Graedel, T. E. 1979. Reduced sulfur emission from the open oceans. Geophys. Res. Lett. 6:329-331.

Granat, L., H. Rodhe, and R. O. Hallberg 1976. The global sulfur cycle. In SCOPE 7: Nitrogen, Phosphorus, and Sulfur--Global Cycles (B. H. Svensson and R. Söderlund, eds.), Stockholm:Royal Swedish Academy of Sciences, 90-110.

Hansen, M. H., K. Ingvorsen, and B. B. Joergensen. 1978. Mechanisms of hydrogen sulfide release from coastal marine sediments to the atmosphere. Limnol. Oceanogr. 23:68-76.

Herrmann, J., W. and Jaeschke. 1984. Measurements of H_2S and SO_2 over the Atlantic Ocean. J. Atmos. Chemistry 1:111-123.

Hobbs, P. V., J. P. Tuell, D. A. Hegg, L. F. Radke, and M. W. Eltgroth. 1982. Particles and gases in the emission from the 1980-81 volcanic eruptions of Mt. St. Helens. J. Geophys. Res. 87:11,062-11,086.

Ingvorsen, K., and B. B. Joergensen. 1982. Seasonal variation in H_2S emission to the atmosphere from intertidal sediments in Denmark. Atmos. Environ. 16:855-865.

Ivanov, M. V., and J. R. Freney. 1983. The Global Biogeochemical Sulfur Cycle. New York:Wiley, 470 p.

Jaeschke, W., H. Claude, and J. Herrmann. 1980. Sources and sinks of atmospheric H_2S. J. Geophys. Res. 85:5639-5644.

Jenkins, W. J. 1982. Oxygen utilization rates in North Atlantic subtropical gyre and primary production in oligotrophic systems. Nature 300: 246-248.

Johnson, J. E. 1981. The lifetime of carbonyl sulfide in the troposphere. Geophys. Res. Lett. 8:938-940.

Kadota, H. and Y. Ishida. 1972. Production of volatile sulfur compounds by microorganisms. Ann. Rev. Microbiol. 26:127-138.

Kellogg, W. W., R. D. Cadle, E. R. Allen, A. L. Lazrus, and E. A. Martell. 1972. The sulfur cycle. Science 175:587-596.

Kotra, J. P., D. L. Finnegan, W. H. Zoller, M. A. Hart, and J. L. Moyers. 1983. El Chichon: Composition of plume gases and particles. Science 222:1018-1021.

Kritz, M. A. 1982. Exchange of sulfur between the free troposphere, marine boundary layer, and the sea surface. J. Geophys. Res. 87:8795-8803.

Krueger, A. J. 1983. Sighting of El Chichon sulfur dioxide clouds with the Nimbus 7 total ozone mapping spectrometer. Science 220:1377-79.

Lein, A. Y. 1983. The sulphur cycle in the lithosphere, part II: Cycling. In The Global Biogeochemical Sulfur Cycle (M. V. Ivanov and G. R. Freney, eds.), New York: Wiley, 95-127.

Liss, P. S. 1983. Gas transfer: experiments and geochemical implications. In Air-Sea Exchange of Gases and Particles (P. S. Liss and W. G. N. Slinn, eds.) Boston: Reidel, 241-298.

Lovelock, J. E. 1974. CS_2 and the natural sulphur cycle. Nature 248:625-626.

Maenhaut, W., H. Raemdonck, A. Selen, R. Van Grieken, and J. W. Winchester. 1983. Characterization of the atmospheric aerosol over the eastern equatorial Pacific. J. Geophys. Res. 88:5353-5364.

Maroulis, P. J., A. L. Torres, A. B. Goldberg, and A. R. Bandy. 1980.
    Atmospheric SO_2 measurements on Project Gametag. J. Geophys. Res.
    85:7345-7349.
McDonald, R. L., C. K. Unni, and R. A. Duce. 1982. Estimation of atmo-
    spheric sea salt dry deposition: Wind speed and particle size depen-
    dence. J. Geophys. Res. 87:1246-1250.
McElroy, M. B., S. C. Wofsy, and N. D. Sze, N.D. 1980. Photochemical
    sources for atmospheric H_2S. Atmos. Environ.14:159-163.
Möller, D. 1984. Estimation of the global man-made sulphur emission.
    Atmos. Environ. 18:19-27.
Nguyen, B. C., C. Bergeret, and G. Lambert. 1984. Exchange rates of di-
    methylsulfide between ocean and atmosphere. In Gas Transfer at Water
    Surfaces (W. Brutsaert and G. H. Jirka, eds.), Dordrecht, Holland:
    Reidel.
Petrenchuk, O. P. 1980. On the budget of sea salts and sulfur in the
    atmosphere. J. Geophys. Res. 85:7439-7444.
Radke, L. F. 1982. Sulphur and sulphate from Mt. Erebus. Nature 299:710-
    712.
Rasmussen, R. A. 1974. Emission of biogenic hydrogen sulfide. Tellus
    26:254-260.
Rasmussen, R. A., M. A. K. Khalil, R. W. Dalluge, S. A. Penkett, and B.
    Jones. 1982a. Carbonyl sulfide and carbon disulfide from the erup-
    tions of Mount St. Helens. Science 215:665-667.
Rasmussen, R. A., M. A. K. Khalil, and S. D. Hoyt. 1982b. The oceanic
    source of carbonyl sulfide (OCS). Atmos. Environ. 16:1591-1594.
Revsbech, N. P., B. B. Joergensen, H. T. Blackburn, and Y. Cohen. 1983.
    Microelectrode studies of the photosynthesis and O_2, H_2S, and pH
    profiles of a microbial mat. Limnol. Oceanogr. 28:1062-1074.
Rodhe, H., and I. Isaksen. 1980. Global distribution of sulfur compounds
    in the troposphere estimated in a height/latitude transport model. J.
    Geophys. Res. 85:7401-7409.
Ryaboshapko, A. G. 1983. The atmospheric sulfur cycle. In The Global
    Biogeochemical Sulfur Cycle (M. V. Ivanov and G. R. Freney, eds.),
    New York:Wiley, 203-296.
Seiler, W., and P. J. Crutzen. 1980. Estimates of gross and net fluxes of
    carbon between the biosphere and the atmosphere from biomass burning.
    Climatic Change 2:207-247.
Shulenberger, E., and J. L. Reid. 1981. The Pacific shallow oxygen maxi-
    mum, deep chlorophyll maximum, and primary productivity, recon-
    sidered. Deep-Sea Res. 28:901-919.
Slatt, B. J., D. F. S. Natusch, J. M. Prospero, and D. L. Savoie. 1978.
    Hydrogen sulfide in the atmosphere of the northern equatorial Atlan-
    tic Ocean and its relation to the global sulfur cycle. Atmos. Envi-
    ron. 12:981-991.
Steudler, P. A., and B. J. Peterson. 1984. Contribution of the sulfur from
    salt marshes to the global sulfur cycle. Nature 311:455-457.
Stith, J. L., P. V. Hobbs, and L. F. Radke. 1978. Airborne particle and
    gas measurements in the emissions from six volcanoes. J. Geophys.
    Res. 83:4009-4017.
Turner, S. M. and Liss, P. S. 1985. Measurement of various sulphur
    gases in a coastal marine environment. J. Atmos. Chem. (in press).

Vairavamurthy, A., M. O. Andreae, and R. L. Iverson. 1985. Biosynthesis of dimethylsulfide and dimethylpropiothetin by Hymenomonas carterae in relation to sulfur source and salinity variations. Limnol. Oceanogr. 30:(in press).

Varhelyi, G., and G. Gravenhorst. 1983. Production rate of airborne sea-salt sulfur deduced from chemical analysis of marine aerosols and precipitation. J. Geophys. Res. 88:6737-6751.

Von Damm, K. L. 1983. Chemistry of submarine hydrothermal solutions at 21° North, East Pacific Rise, and Guaymas Basin, Gulf of California. Ph. D. Dissertation, Massachusetts Institute of Technology, Cambridge, Mass., 241 p.

Winer, A. M., R. Atkinson, and J. N. Pitts, Jr. 1984. Gaseous nitrate radical: Possible nighttime atmospheric sink for biogenic organic compounds. Science 224:156-158.

Winner, W. E., C. L. Smith, G. W. Koch, H. A. Mooney, J. D. Bewley, and H. R. Krouse. 1981. Rates of emission of H_2S from plants and patterns of stable sulphur isotope fractionation. Nature 289:672-673.

Zinder, S. H., W. N. Doemel, and T. D. Brock. 1977. Production of volatiles sulfur compounds during the decomposition of algal mats. Appl. Environ. Microbiol 34:859-860.

2. THE EMISSION OF NITROGEN TO THE REMOTE ATMOSPHERE: BACKGROUND PAPER

Ian E. Galbally
CSIRO Division of Atmospheric Research
Private Bag No. 1
Mordialloc, Victoria, 3195 Australia

2.1. INTRODUCTION

Two questions spring to mind when one thinks of the emission of nitrogen compounds to the atmosphere of remote areas: Firstly, Are there any data on these fluxes? Secondly, after some reflection, Is there any area that can be described as remote for the nitrogen cycle? This review attempts to answer these questions.

The four basic processes that lead to the introduction of fixed nitrogen compounds into the atmosphere are (1) nitrogen transformations during nitrification and denitrification, (2) ammonia volatilization, (3) combustion, and (4) lightning. The degree to which each process is active at a specific location depends on the climate and the biological activity in that region--or, in modern parlance, on the type of ecosystem. There is a need in quantitative biogeochemical studies to define rigorously these ecosystems and the areas of the world that the ecosystems represent. However, although such researchers as A. Rodin, N. I. Bazilevich, H. Lieth, and R. H. Wittaker have made attempts at global biological mapping during the last few decades, the most recent studies by Hummel and Reck (1979), Olson and Watts (1982), and Matthews (1983) still disagree considerably. These disagreements seem to be caused partly by different definitions but also partly by the data.

A further complication arises because the ecosystems where nitrogen emissions have been measured have not yet been clearly defined. Only limited information on location, etc., can be deduced from the original papers. Furthermore, in the absence of comprehensive emission data, some choices have to be made concerning how many categories of ecosystems can be used that will allow the limited available emission data to be extrapolated to as wide an area as is reasonable without introducing major errors into the analyses because of the widely divergent ecosystems incorporated into the one category. To achieve this, I have selected a grossly simplified classification where, although the various categories could not be described as ecosystems, they do have value for classifying nitrogen emissions (Table 2-1). In this paper I estimate or comment on the fixed-nitrogen emissions from each category in Table 2-1.

J. N. Galloway et al. (eds.), The Biogeochemical Cycling of Sulfur and Nitrogen in the Remote Atmosphere, 27–53.
© 1985 by D. Reidel Publishing Company.

Table 2-1. Major Ecosystem Subdivisions Grouped by Categories Suitable
    for Nitrogen-Emission Estimates.

Categories for N-Emission Estimates	Major Ecosystem Subdivisions	
	Map Symbol	Description
Tundra	M	Arctic/alpine tundra, mossy bog
Ice	V	Ice
Woodland, shrub land and grass- land	C	Xeromorphic forest/woodland
	D	Evergreen broad-leaved sclerophyllous woodland
	D	Evergreen needleleaved woodland
	F	Tropical/subtropical drought-deciduous woodland
	G	Cold-deciduous woodland
	H	Evergreen broad-leaved shrub land/thicket, evergreen dwarf shrub land
	I	Evergreen needleleaved or microphyllous shrub land/thicket
	J	Drought-deciduous shrub land/thicket
	K	Cold-deciduous subalpine/subpolar shrub land, cold-deciduous dwarf shrub land
	L	Xeromorphic shrub land/dwarf shrub land
	N	Tall/medium/short grassland with 10%-40% woody tree cover
	O	Tall/medium/short grassland with <10% woody tree cover or tuft-plant cover
	P	Tall/medium/short grassland with shrub cover
	Q	Tall grassland, no woody cover
	R	Medium grassland, no woody cover
	S	Meadow, short grassland, no woody cover
Cultivated lands	W	Cultivation
Forests	1	Tropical evergreen rain forest, mangrove forest
	2	Tropical/subtropical evergreen seasonal broad-leaved forest
	3	Subtropical evergreen rain forest
	4	Temperate/subpolar evergreen rain forest
	5	Temperate evergreen seasonal broad-leaved forest, summer rain
	6	Evergreen broad-leaved sclerophyllous forest, winter rain
	7	Tropical/subtropical evergreen needle-leaved forest
	8	Temperate/subpolar evergreen needle-leaved forest
	9	Tropical/subtropical drought-deciduous forest
	A	Cold-deciduous forest, with evergreens
	B	Cold-deciduous forest, without evergreens
Deserts	U	Desert

Source: Data for major ecosystem subdivisions from Matthews (1983).

2.2. NITRIFICATION AND DENITRIFICATION

For the purposes of this discussion, nitrification and denitrification
are microbial processes in soil or water that can lead to the production
of nitrous oxide, N_2O; nitric oxide, NO; and perhaps nitrogen dioxide,
NO_2. NO and NO_2 ($NO_x \equiv NO + NO_2$) are probably also produced through nonbiological processes, called chemodenitrification, following the biological
production of nitrite, NO_2^-.
   The accepted sequence for denitrification is

$$NO_3^- \longrightarrow NO_2^- \longrightarrow NO \longrightarrow N_2O \longrightarrow N_2 \qquad\qquad (2\text{-}1)$$

(Payne 1981). Denitrification is an anaerobic process carried out by
bacteria. At least 146 strains that are capable of complete denitrification (production of molecular nitrogen, N_2, from NO_3^-) have been found in
soil and freshwater samples from around the world. Many other bacteria
are capable of catalyzing one or more steps in the denitrification
sequence (Ingraham 1981). Denitrification is repressed by oxygen, O_2.
Hence, denitrification is often associated with soil-water contents above
60% water-holding capacity (Knowles 1981). Most denitrifying bacteria
are heterotrophs and require readily oxidizable organic carbon. For NO_3^-
concentrations above 40 μg N/g, the denitrification reaction appears to
be zero order with respect to NO_3^-. However, nitrate levels in unfertilized soils are generally lower. Laboratory and field studies have shown
that, at very low nitrate levels (less than 1 μg N/g), N_2O can be consumed by soils (Blackmer and Bremner 1976, Freney et al. 1978, Cicerone
et al. 1978, and Ryden 1981).
   The rate of denitrification is temperature dependent and roughly
doubles for every 10° C rise (i.e., Q_{10} = 2). The maximum rate occurs
at around 60° to 75° C and measurable denitrification occurs even from 0°
to 5° C (McKenney et al. 1980). The optimum pH appears to range from 7.0
to 8.0. The presence of sulfide inhibits denitrification. (For details
of these processes, see Bryan 1981, Knowles 1981, Fillery 1983, and Rosswall 1982). Thus, denitrification appears to flourish in warm, moist
soils containing both NO_3^- and available organic carbon. Field experiments show that denitrification is, in fact, an episodic event.
   There is no specific laboratory evidence concerning the ratio of N_2O
to NO emissions during denitrification. The works of Johansson and Galbally (1984), McKenney et al. (1982), and those cited earlier suggest
that, because NO and N_2O can be simultaneously produced and consumed by
the soil microorganisms, the ratio of N_2O to NO depends on physical
transfer processes from the site of the denitrification to the atmosphere
as well as on such factors as the relative rates of production and
consumption of NO and N_2O. In suitable conditions, essentially all the
NO_3^- in soils can be consumed, producing NO, N_2O, and N_2. The field
evidence suggests that N_2 is the primary product.
   Nitrification involves the biological oxidation of fixed nitrogen.
The most common form involves the oxidation of ammonium, NH_4^+, to NO_3^-
with NO_2^- as an intermediate (Bremner and Blackmer 1981). Separate
bacteria oxidize NH_4^+ to NO_2^- and NO_2^- to NO_3^-. These bacteria are
generally chemoautotropic and require only CO_2, H_2O, O_2, and either NH_4^+
or NO_2^- to grow.

Ritchie and Nicholas (1972) have reported on nitrogen transformations within the autotrophic nitrifying bacterium **Nitrosomonas** that
converts NH_4^+ to NO_2^- and is also able to reduce NO_2^-. Their study shows
that N_2O release is an alternative terminal pathway in the nitrifying
process. They propose NO as an intermediate prior to NO_2^- production but
subsequent to the branch leading to N_2O production. Ritchie and Nicholas
suggest that this same bacterium, when reducing NO_2^-, produces NO first
as an intermediate and subsequently N_2O.

It is debatable whether or not NO is an obligatory intermediate
(presumably enzyme bound) in the nitrification sequence because an
enzyme-catalyzed conversion of NO to NO_2^- has not yet been demonstrated
(Hooper and Terry 1979, Hooper 1978). (See Verstraete 1981, Bremner and
Blackmer 1981, and Rosswall 1982 for reviews of the properties of nitrifying bacteria.)

One laboratory study has found increased NO release relative to NO_2^-
production for a soil **Nitrosomonas** operating at reduced O_2 levels of 0.5%
versus 21.0% (Lipschultz et al. 1981). This study also shows that the
mole ratio of NO/N_2O produced by the bacterium vary from ~ 5 at 0.5% O_2
level to ~1 at 21% O_2. No other evidence of environmental factors influencing NO release by nitrifying bacteria is available.

Only a small fraction of the nitrogen oxidized during nitrification
is lost as NO or N_2O. In laboratory experiments, Johansson and Galbally
(1984) measured NO production from soil that corresponds to about 2% of
the rate of nitrification in aerobic conditions (subject to several
qualifications). Lipschultz et al. (1981) report that between 0.3% and
10% of the NH_4^+ oxidized in an aqueous growth medium is converted to NO
or N_2O. Nitrification is found in any well-aerated soil in the presence
of NH_4^+ or urea, $(NH_2)_2CO$, except perhaps in a very acidic soil that has
a pH ≤ 4 (Rosswall 1982).

Part of the fixed nitrogen transformed during denitrification or
nitrification can be lost via an indirect pathway. Depending on the
bacteria present, either process can lead to the temporary accumulation
of NO_2^- in soil. This NO_2^- can decompose abiotically, yielding NO or
NO_2, or can react with soil lignins forming methyl nitrite, CH_3NO_2. These
processes are generally described as ''chemodenitrification'' (Broadbent
and Clark 1965, Allison 1965). Although many laboratory studies of these
processes have been published (e.g., Nelson and Bremner 1970, Bulla et
al. 1970, Steen and Stojanovic 1971, Nommik and Thorin 1972, van Cleemput
et al. 1976, Moraghan and Buresh 1977), the only general agreement about
the mechanisms or environmental factors influencing the rate of NO production (Chalk and Smith 1983) is that these processes should be substantially favored in acidic soils.

2.3. THE PRODUCTION OF FIXED NITROGEN DURING COMBUSTION AND LIGHTNING

Both lightning and combustion can produce nitrogen oxides in the atmosphere. These gases are produced through the high-temperature dissociation of N_2 in the presence of ''excess air'' in combustion or normal air
in lightning through the Zeldovich mechanism,

$$O + N_2 \longrightarrow NO + N \qquad\qquad\qquad (2-2)$$

$$N + O_2 \longrightarrow NO + O, \hspace{5cm} (2\text{-}3)$$

where the first reaction is rate determining. The rate of NO formation depends on temperature, the presence of water vapor, and molecular hydrogen (Mulcahy 1973). The overall equation for this process, which is profoundly affected by temperature, is

$$N_2 + O_2 = 2NO \hspace{0.5cm} (\Delta \bar{H}^O = + 181 \text{ kJ/mole}). \hspace{2cm} (2\text{-}4)$$

The equilibrium constant is around 10^{-3} at 2500 K and 10^{-30} at 300 K. The rate at which this equilibrium is reached also varies strongly with temperature, taking around 10^{-3} sec at 2500 K and more than doubling this value for each 100 K of temperature decrease. Thus, for air that is rapidly heated and cooled, as in a flame or lightning stroke, the NO formed at a high temperature can be ''frozen'' within the air. This happens because, with the long equilibrium time required at lower temperatures, the relevant equilibrium is never attained (Campbell 1977). This is an important process in the formation of atmospheric NO_x.

Hydrogen cyanide, HCN, can also be formed during combustion in the gas phase by reactions including

$$CH + N_2 \longrightarrow HCN + N. \hspace{4cm} (2\text{-}5)$$

However, thermodynamic calculations on CH_4 flames indicate that this gas-phase production is negligible. The concentration of HCN is 10^{-6} to 10^{-9} times less than that of NO at 2000° C and the difference is even larger at lower temperatures (unpublished data available from Dr. M. Y. Smith, CSIRO, North Ryde, NSW 2113, Australia).

The other source of the fixed nitrogen formed during combustion is the nitrogen within the compounds in fuel. Manskaya and Drozdova (1968), in discussing peat, coal, etc., indicate that the N in these compounds may occur in end groups, in open chains, and in heterocyclic rings. During pyrolysis or combustion, this nitrogen can pass into the gas phase as decomposition or combustion products, such as NH_3, NO, N_2O, N_2, or even HCN. The latter could be formed from the pyrolysis of nitrogen-containing heterocyclic rings. Fuel nitrogen is liberated at all temperatures at which smoldering and combustion occur. The nitrogen content of different fuels varies considerably. Typical values include 0.2%–1% for dead-plant matter, 0.5%–2% for peat, 1%–2% for coal, and 1%–2% for fuel oil.

Logan (1983) has reviewed the partitioning of fuel nitrogen between various gaseous products. In one laboratory study of the low-temperature combustion (1270 K) of vegetation samples, 23% of the fuel N is released as NO_2 (Clements and MacMahon 1980). This is, as Logan (1983) points out, consistent with conversion efficiencies for fuel N in the burning of coal. However, these results may not be applicable to large-scale fires. Clements and McMahon (1980) have reported burning 9 mg of fine-ground fuel in a furnace with a maximum heating rate of 80° C/min from 100° C to 1000° C with a minimum oxygen-flow rate of approximately 1.2 mg/min. Thus, on first appearance, it seems that their range of experimental conditions do not cover oxygen-limited fires. Typical large-scale fires

(Crutzen et al. 1979) show $CO:CO_2$ ratios of 1:10, which suggests that
oxygen limitation is a significant factor in these fires. The physical/
chemical influences of the differences, between Clements and MacMahon's
(1980) findings for combustion and those for large-scale fires, on the
conversion of fuel N to NO_x are unknown. However, it is quite possible
that they could have a profound effect.

Evans et al. (1977) have reported data on CO_2 and NO_x produced by
large-scale fires in Australia. Table 2-2 shows data from the literature
on the nitrogen contents of fire fuels from these areas. These nitrogen
contents are converted to N:C ratios assuming that wood and forest litter
are 45% C and that dry grass is 40% C (see references in Table 2-2). By
comparing the ratio of NO_x-N to excess CO_2-C (above ambient atmospheric
CO_2) in the smoke to the ratio of N:C in the fuel, I calculated how
efficiently fuel N is converted to NO_x in these large-scale fires.
Table 2-3 shows that the efficiency of conversion is 5% to 19% with a
median value around 9%.

Delany and Wartburg (1983) have found similar NO_x-N:CO_2-C ratios in
smoke plumes from the dry-season burning of the savanna region of central
Brazil. Unfortunately, the N content of the litter from this area is not
readily available. Although the work of Medina (1982) on Venezuelan
grasslands suggests a nitrogen content of 0.3% in the burnt material, the
litter fall from trees in these regions can contain much higher N content
(Wetselaar 1980, Braun Wilke 1982) and can contribute 50% of the total
litter load. If we assume that a total N content of 0.6% is appropriate
for the litter of the Brazilian Cerrado, then the data of Delaney and
Wartburg correspond to a 12% conversion efficiency of fuel N to NO_x.

These results from large-scale fires are about a factor of 2 lower
than the laboratory results of Clements and McMahon (1980). For reasons
discussed earlier, I have used the values derived here, which are based
on large-scale fires, to calculate the atmospheric source of NO_x from

Table 2-2. Fuel-N Contents for Various Regions in Australia.

Location and Fuel	Fuel N (% of dry matter)	Reference
Western Australia		
Forest litter	0.4 - 0.8	Hatch 1955
		Vines et al. 1971
Whole above-ground forest biomass	0.1 - 0.5 (0.3 av)	Hingston & Raison 1982
Northern Australia		
Tropical grassland	0.7	Wetselaar & Hutton 1963
		Wetselaar 1980
Tropical forest		
Brigalow	0.5	Moore et al. 1967
Arid shrub	0.6 - 1.6	Burrows 1972
Mangrove	1.3	Boto & Bunt 1982
Mangrove litter	0.3	Boto & Bunt 1982

Table 2-3. Estimates of the Conversion Efficiency of Fuel N to NO_x in Large-scale Fires in Australia.

Fire Location and Type	Fuel N:C Ratio (%)	Smoke NO_x-N:CO_2-C Ratio (%)	N Conversion Efficiency (%)
Western Australia			
Prescribed forest burn	1.3	0.09	7
Clearing burn:			
felled forest	0.3	0.04	13
Northern Australia			
Grass fires	1.8	0.16	9
Prescribed burn:			
swamp flats	0.7–2.9	0.13	5–19

Source: Smoke data from Evans et al. (1977).

biomass burning. The NO_x-N emitted is calculated as the product of the biomass burnt times the average nitrogen content of the biomass times the fuel-N conversion efficiency.

NH$_3$ and N_2O are also produced during combustion. Unfortunately, no quantitative information is available on NH$_3$ production but some is available on N_2O production. Crutzen et al. (1979) have reported measurements of the N_2O:CO_2 ratio produced in two large fires in the United States. In a recent study of the products of fire on the Brazilian Cerrado, Crutzen and his colleagues (1985) suggest a tentative value of $2.5 \cdot 10^{-4}$ for the N_2O:CO_2 ratio from such fires (i.e., ~ $6 \cdot 10^{-4}$ N_2O-N:CO_2-C ratio). I have used this figure in subsequent analyses in this paper to predict the N_2O emissions from biomass burning.

2.4. VOLATILIZATION OF AMMONIA

The process of ammonia volatilization has been described recently by Freney et al. (1983). Since NH$_3$ is a gas at normal atmospheric temperatures and pressures, it may be expected that any NH$_3$ in soils, waters, fertilizers, manures, or exposed parts of plants would quickly volatilize to the atmosphere. However, NH$_3$ is a basic gas that reacts readily with protons, metal ions, and acidic compounds to form ions, compounds, or complexes of varying stability. It can, therefore, be retained in solution or in solid form. Ammonia also has a very strong affinity for water and its reactions in water constrain the rate of volatilization.

The driving force of NH$_3$ volatilization from a moist soil or a solution is normally considered to be the difference in the NH$_3$ partial pressure between that in equilibrium with the liquid phase and that in the ambient atmosphere (Freney et al. 1981a, Denmead et al. 1982). The equilibrium vapor pressure of NH$_3$ is controlled by the NH$_3$ concentration

in the adjacent solution that, in the absence of other ionic species, is
affected by the ammonium ion concentration and the pH. These equilibria
are so strongly temperature dependent that, for a fixed pH and NH_4^+ concentration in solution, the equilibrium vapor pressure of NH_3 rises by
about a factor of 3 for each $10°$ C the temperature increases, i. e.,
Q_{10} = 3 (Farquhar et al. 1980).

NH$_3$ volatilization from soils is affected by (1) the binding of NH_4^+
to soil clays, (2) the soil pH, (3) the soil-buffer capacity, (4) the
presence of $CaCO_3$, (5) the temperature, (6) the evaporation, (7) the
turbulent mixing, and (8) the presence of plants. Most of these processes are well understood, and Dawson (1977) has produced a physical
model for NH_3 volatilization from soils throughout the world.

However, two additional points need to be discussed. The first
concerns the horizontal inhomogeneity of NH_3 emissions from grazed areas.
A considerable fraction of the NH_3 that is volatilized does not come from
a homogeneous "background" soil/litter system. Soil fauna graze on
microflora (primary decomposers) and plant mineral and excrete NH_3,
$(NH_2)_2CO$, and amino acids. This is primarily a below-ground activity
where the excreta are diffusely spread and comparatively isolated from
the atmosphere (Anderson et al. 1981). However, above-ground grazing by
large herbivores presents a different picture (Floate 1981). Plant material is gathered over a wide area and then the excreta are deposited in a
much smaller area. In a well-grazed system, some 40%-80% of N in the
herbage may be consumed of which 50%-75% may be partitioned into urine.
With sheep, this is deposited on 30% of the grazed area and with cattle,
on 15% of the grazed area. In natural ecosystems, the fractional area
"refertilized" may be much smaller. The urine or dung patches that
constitute this return behave quite differently to the areal "average"
and the high concentrations of NH_3 and $(NH_2)_2CO$ promote high pH through
the hydrolysis of $(NH_2)_2CO$ to NH_4^+. Within these patches volatilization
is enhanced through the separate but related effects of high-pH and high-
NH_3 concentrations.

The second factor to be considered in studies of NH_3 volatilization
is the role of plants. Denmead et al. (1976) have found that NH_3 emissions from the surface of the soil of an ungrazed grass-clover pasture
are emitted at rates comparable to those for grazed pastures. However,
most NH_3 released from the soil surface is adsorbed by the plants above
the surface. Denmead et al. (1978) have reported a similar complex
pattern of NH_3 uptake and release in a growing corn crop. The release of
NH_3 from the soil in these cases is presumably caused by the microbial
decomposition of old plant material in the soil (Dawson 1977).

Farquhar et al. (1979, 1980) provide a key to understanding this
process. They infer that there is a partial pressure of NH_3, $p(NH_3)$, in
the intercellular cavities of plants that is presumed to be in equilibrium with NH_3 and NH_4^+ in the cellular fluid. This NH_3 and NH_4^+ results
from a variety of reactions in the plant cells. The equilibrium partial
pressure of NH_3 within the plant is known as the compensation point for
NH_3, γ. This compensation point has a temperature dependency and, for
the four species measured so far, values of 2 to 6 nbar at $20°$ C (Farquhar et al. 1980). The intercellular cavities where this equilibrium
occurs are connected via stomata to the atmosphere. Depending on whether

the partial pressure of NH_3 in the surrounding atmosphere is higher or
lower than the equilibrium concentration in the plant, NH_3 will either
flow into the plant or be released from it.

Similarly NH_3 is probably released at an accelerated rate during
plant senescence when the uptake process ceases and proteins break down
in the senescing leaves, liberating NH_3 (Wetselaar and Farquhar 1980).
Volatile amines may also be liberated by this process (Farquhar et al.
1980). Although HCN has been identified as an emission from sorghum
plants (Franzke and Hume 1945), there is no strong evidence of oxidized
forms of fixed nitrogen being volatilized by plants (Farquhar et al.
1983).

Field observations summarized by Farquhar et al. (1983) suggest NH_3
compensation points of 2 to 10 nbar over various crops with temperatures
from 18° to 24° C. The equilibrium partial pressure of NH_3 over a solution of unchanging pH and unchanging total ammonium content ($NH_3 + NH_4^+$)
depends highly on temperature, just as the γ for plants probably does.
Therefore, ammonia volatilization from plants occurs primarily in hot
conditions with adequate water supply (for stomatal opening) and during
plant senescence.

The molar flux density of NH_3 between plants and the atmosphere, F,
can be expressed as

$$F = \frac{g}{P} \ p(NH_3)_{atmos} - \gamma \ , \qquad\qquad (2-5)$$

where g is the conductance for the diffusion of the gas through the
boundary layer of the leaf and the stomata and P, the atmospheric pressure. When $p(NH_3)_{atmos} > \gamma$, NH_3 is taken up by the plant whereas when
$p(NH_3)_{atmos} < \gamma$, NH_3 is released from the plant to the atmosphere. Similar descriptions of equilibria and fluxes have been reported for soils
and oceans (Dawson 1977, Georgii and Gravenhorst 1977).

2.5. EMISSIONS AND DRY DEPOSITION: CAN THEY BE SEPARATED?

Traditional budget studies and conceptual models of NO_x, NH_3, and N_2O in
the remote atmosphere incorporate emissions and dry deposition as distinct processes that occur simultaneously over (or at) the same surface.
These emissions and dry deposition, when evaluated separately, represent
the gross fluxes at the surface. Sometimes these gross fluxes are of
interest, e.g., from an ecological viewpoint when the emissions and
deposition represent the cycling of nitrogen (or some other element) from
one set of organisms or plants to another set in the same ecosystem.
However, for global or regional budgets, the more appropriate flux would
be the net flux over the surface. The use of gross fluxes could give the
mistaken impression that excessively large amounts of the species in
question were being cycled through the atmosphere.

Generally, the present concept of emissions is one of a constant
flux applicable to the whole area of an ecosystem or group of ecosystems.
This flux either represents a long-term average or varies in some simple
way with the season. The rate of "dry" deposition of a gas is usually

calculated from the product of the concentration of the gas at a reference height above the surface and a deposition velocity chosen, hope-
fully, to be representative of the surface. Both the gas concentration
and the deposition velocity are assumed to be areally representative of
the ecosystem and minimal temporal variations are included in the calculations. These concepts can be combined in many situations to yield net
fluxes and the resulting net fluxes are plainly more appropriate than the
gross fluxes for atmospheric modeling and budget studies.

   Another relevant concept is that of an "equilibrium" concentration
appropriate to a particular surface where, if the equilibrium concentration prevails, there is zero net flux from the surface. These equilibrium concentrations are calculated as the quotient of the emission rate
over the deposition velocity. The equilibrium concentration may be
compared with actual measurements of concentration in the air above a
surface for use as a guide in determining the direction of the net flux.
When the atmospheric concentration is greater than the equilibrium concentration, then net uptake is likely to occur. When the concentrations
are vice versa, net release is likely to occur. The concept of equilibrium concentrations provides a method by which, through examining the
available atmospheric concentration measurements in the planetary-
boundary layer, some inference about the direction of the net flux (and
through more complicated argument possibly its magnitude) can be
obtained.

   The odd nitrogen oxides, NO and NO_2, have a complicated pattern of
exchange. Johansson and Galbally (1984) demonstrate that soil bacteria
functioning in aerobic conditions both release and consume NO, depending
on the NO concentration in the soil atmosphere. The equilibrium concentration in the soil atmosphere at which there is no net exchange varies
from 200 to 400 ppbv. These authors suggest that denitrifying bacteria
behave similarly. Other studies from the ambient atmosphere (Rogers et
al. 1979, Judeikis and Wren 1978, Galbally and Roy 1981, 1983a 1983b)
show that NO and NO_2 are taken up by soil and vegetation. NO_2 is taken
up some two to three times more rapidly than NO for comparative concentrations of the two gases. More recently Slemr and Seiler (1983) and
Johansson and Granat (1984) have shown that NO_2 as well as NO is emitted
from soils. In the presence of vegetation cover, no NO_2 but a substantial
fraction of the NO released reaches the atmosphere. Thus, for NO and NO_2
we have a complex picture: At the biological level and subsequently,
both uptake and release can occur. The net flux of NO or NO_2 from a
surface depends on soil microbial processes, the presence or absence of
plant cover, and the atmospheric concentration of the gas, either NO or
NO_2, being considered.

   Galbally and Roy (1981, 1983a, 1983b) discuss two types of this
release/uptake pattern, which they have observed. The first form involves NO released from emission "hot spots" and its concurrent uptake
at other sites around the field where uptake exceeds emissions. The
second form involves the net release of NO, its oxidation in the air by
O_3 to form NO_2, and the net uptake of NO_2. Slemr and Seiler (1983) and
Johansson and Granat (1984) have shown examples of the NO_x exchange
varying from net emission to net uptake at specific sites.

   This situation is somewhat clarified by introducing a compensation
point, or equilibrium concentration, at which the net exchange of the gas

is zero. For NO this equilibrium concentration appears to be around 3 to 10 ppb (Galbally and Roy 1983b) and when ambient NO concentrations are lower, net emission occurs. In fact, because NO concentrations in remote areas are generally much lower (see the review of data in Galbally and Roy 1983a), we can presume that net NO emission generally occurs. The experimental data have shown no evidence of the net emission of NO_2 in the presence of vegetative cover (Galbally and Roy 1978), but they do show evidence of the net uptake of NO_2 (Galbally and Roy 1983a, Slemr and Seiler 1983, Johansson and Granat 1984). Thus, I assumed no net NO_2 emission in the following analyses.

The situation with NH_3 is somewhat similar. Studies of plants (Farquhar et al. 1979, 1980), land surfaces (Dawson 1977), and oceans (Georgii and Gravenhorst 1977) have shown that whether the surface acts as a source or sink for atmospheric NH_3 depends on whether the partial pressure of NH_3 in the atmosphere exceeds or is less than some equilibrium partial pressure of NH_3 for the underlying surface. Indeed, there are probably adjoining regions where ·the underlying surfaces are alternately ammonia-rich and -poor. A sequence of emissions by the ammonia-rich underlying surface, short-range transport by the atmosphere, and uptake by the surface in the adjoining region make up part of the atmospheric ammonia cycle.

Because atmospheric concentrations of NH_3 at remote sites are so varied (see a review of data in Galbally and Roy 1983a), it is impossible from the available concentration data and knowledge of surface processes to determine whether various pristine ecosystems are acting as net sources or sinks for atmospheric NH_3.

N_2O is an intermediate in the process of denitrification. Therefore, because the atmosphere is a large reservoir of N_2O, it would not be surprising if soil microorganisms were drawing upon this source when other forms of oxidized N are in short supply. However, because the consumption of N_2O is inhibited by O_2 and NO_3^-, ''even very wet soils are more likely to be sources than sinks as long as the soil surface has an oxidized layer and mineralizable organic nitrogen is present'' (Freney et al. 1978). Ryden (1981) has given the only field measurements of persistent N_2O uptake available, and these occur at high soil-moisture content (> 70% of field capacity) when NO_3^- is depleted to ≤1 µg N/g. The addition of nitrate fertilizer stimulates N_2O production in this soil. Cicerone et al. (1978), in an exploratory study of N_2O sources and sinks, have reported that, on very wet, mostly shaded grassy areas, uptake of N_2O is found on 8 of 60 days examined even though N_2O is produced on the same days at drier, sunnier test spots only 25 m away.

None of the other 17 studies of N_2O emissions from soil discussed in this paper have reported observing N_2O uptake in spite of the fact that many of those studies involve measurements over several sites for a year or more. Therefore, although N_2O can be both taken up and released at the earth's surface, emission prevails both in space and time and no special qualifications of emission estimates are necessary.

2.6. EMISSIONS OF NO_x FROM SOILS

In this section I list the available measurements of NO_x produced within the various ecosystems described in Table 2-1. No values are available for either a tropical or a boreal forest or for tropical woodlands, shrub lands, or grasslands. The other ecosystems for which no values have yet been reported are tundra and alpine meadows, deserts and shrublands, or swamps and meadows.

2.6.1. Temperate Forests

Johansson (1984) has measured NO fluxes in Sweden from unfertilized forest soils during the summer of $3 \cdot 10^{-10}$ g N/($m^2 \cdot$ s). Based on a 150-day growing season, this would give a yearly flux of 0.004 g N/($m^2 \cdot$ yr).

2.6.2. Temperate Woodlands, Shrub Lands, and Grasslands

Galbally and Roy (1978) have presented three measurements of NO fluxes over grasslands of 0.6 to $2.6 \cdot 10^{-9}$ g N/($m^2 \cdot$ s), which over a year would give an average of 0.02 to 0.08 g N/($m^2 \cdot$ yr). Slemr and Seiler (1983) have measured NO_x losses of 0.03 g N/($m^2 \cdot$ yr) for bare soil in a grass meadow in West Germany. When the grass cover is intact, there is a net uptake of -0.006 g N/($m^2 \cdot$ yr).

2.6.3. Agricultural Lands

The NO_x emissions (both NO and NO_2) from unfertilized agricultural land varied from 0.02 g N/($m^2 \cdot$ yr) for unfertilized barley in Sweden (Johansson and Granat 1984) to 0.24 g N/($m^2 \cdot$ yr) for fallow land in Spain (Slemr and Seiler 1983). Galbally and Roy (1978, 1981, 1983a, 1983b) have found NO fluxes from three pastures in southeastern Australia that give an average flux of 0.1 g N/($m^2 \cdot$ yr) for measurements covering months in autumn, winter, and spring.

NO_x loss rates from fertilized soils have also been measured by Johansson and Granat (1984) and Slemr and Seiler (1983). NO emissions from a grass ley fertilized with 20 g Ca(NO_3)/m^2 are 0.06 g N/($m^2 \cdot$ yr), which suggests a 0.3% loss rate. Slemr and Seiler (1983) reported a 0.2% loss rate for nitrate fertilizer, 1% for ammonium fertilizer, and roughly 1% for urea in the presence of vegetation. Higher loss rates are recorded over bare soils.

In light of background rates and fertilizer usage, the NO_x emissions from agricultural land seem to range from 0.04 g to 0.26 g N/($m^2 \cdot$ yr).

2.7. EMISSIONS OF N_2O FROM SOILS

In this section I briefly list the available measurements on N_2O produced within the ecosystems described in Table 2-1. No measurements are available for a boreal forest.

2.7.1. Tropical Forests

Keller et al. (1983) have measured N_2O fluxes of 0.2 g N/($m^2 \cdot$ yr) in an Amazonian rain forest. This is consistent with Bentley et al.'s (1982) estimate of 1 g N/($m^2 \cdot$ yr) lost by denitrification ($N_2 + N_2O + NO$) in the Amazon basin.

2.7.2. Temperate Forests

Duxbury et al. (1982) have reported emission measurements at a forested ''mineral soil'' site in New York of 0.09 g N/(m^2 · yr). Keller et al. (1983) have measured N_2O in the Hubbard Brook Forest of New Hampshire where they found a flux of 0.007 g N/(m^2 · yr).

2.7.3. Tropical Woodlands, Shrub Lands, and Grasslands

Duxbury et al. (1982) have reported measurements at an ''organic soil'' site in the Florida Everglades Conservation Area that result in emissions of about 0.1 g N/(m^2 · yr). Yates et al. (1982) suggest that the total denitrification rate (N_2 + N_2O + NO) in Latin American savannas is < 0.1 g N/(m^2 · yr), which would imply average N_2O fluxes much smaller than Duxbury et al. (1982) have reported.

2.7.4. Temperate Woodlands, Shrub Lands, and Grasslands

Duxbury et al. (1982) have presented data for fluxes of 0.09 g to 0.17 g N/(m^2 · yr) for an unmanaged timothy-grass, mixed-weed stand. Seiler and Conrad (1981) have found N_2O loss rates of 0.004 g to 0.1 g N/(m^2 · yr) with a median of 0.02 g N/(m^2 · yr) for ''natural, not agriculturally used,'' grass-covered soils. Denmead et al. (1979) have measured N_2O emissions from an uncultivated grass sward of 5 · 10^{-5} g N/(m^2 · day) to 2 · 10^{-2} g N/(m^2 · day). Although Denmead et al.'s (1979) data do not cover a complete year, I calculated a release rate of ~ 0.1 g N/(m^2 · yr) by taking the springtime rate as applicable for 100 days of the year. Ryden (1981) has measured N_2O emissions from unfertilized grassland as a control and find N_2O uptake by the soil. This sink rate is equivalent to an uptake of ~ 0.02 g N/(m^2 · yr). No other researchers have reported such a sink although it is quite consistent with laboratory results at extremely low nitrate concentrations (< 1 μg N/g).

2.7.5. Tundra and Alpine Meadows

No measurements of N_2O emissions for tundra or alpine meadows are available. However, total denitrification (N_2 + N_2O + NO) at a wet meadow at Barrow, Alaska, of 0.003 g N/(m^2 · yr) is reported by Van Cleve and Alexander (1981). The N_2O flux would be a small fraction of this amount.

2.7.6. Deserts and Shrub Lands

No measurements of N_2O emissions for desert or shrub lands are available. Total denitrification (N_2 + N_2O + NO) in arid ecosystems is estimated to almost equal the N fixed during the season (Skujins 1981) and average fixation rates (in cryptogamic crusts) vary from 0.05 g N/(m^2 · yr) for the Mohave Desert to 1.3 g N/(m^2 · yr) for the Great Basin Desert. N_2O emission rates might be an order of magnitude smaller than these rates.

2.7.7. Swamps and Wetlands

Smith et al. (1982) have measured a N_2O loss rate in a salt marsh of the
Mississippi deltaic plain of 0.05 g $N/(m^2 \cdot yr)$.

2.7.8. Agricultural Lands

I included fluxes from agricultural ecosystems here because, even in the
most ''remote'' regions, some agricultural activity is present. In evalu-
ating the N_2O flux from agricultural land, two factors have to be considered: Firstly, N_2O emitted in the absence of nitrogenous fertilizers
and, secondly, N_2O emitted as a direct result of fertilizer enhancement
of soil inorganic nitrogen.
      Roy (1979) has measured N_2O fluxes from an Australian pasture of
0.06 g $N/(m^2 \cdot yr)$. Conrad et al.'s (1983) measurements from ungrazed
grassland cropped for grass give N_2O fluxes of 0.02 g $N/(m^2 \cdot yr)$.
Bremner et al. (1980), in a study of six agricultural soils from Iowa,
have found an average of 0.2 g $N/(m^2 \cdot yr)$ is lost from unfertilized
soils planted with soy beans. Breitenbeck et al. (1980) have reported
emissions of 0.04 g $N/(m^2 \cdot yr)$ for unfertilized agricultural soil (estimated using the seasonal variability of N_2O emission from Bremner et al.
1980). Duxbury et al. (1982) have found N_2O fluxes from unfertilized
agricultural soils (both mineral and organic) in New York and Florida
ranging from 0.2 g to 16.5 g $N/(m^2 \cdot yr)$ with the median value from 8
trials lying between 1.6 g and 4.8 g $N/(m^2 \cdot yr)$. The higher flux values
are for warm, wet organic soils where the nitrogen mineralization is
estimated to be 60–120 g $N/(m^2 \cdot yr)$, most of which is probably lost
through denitrification.
      The loss of fertilizer nitrogen as N_2O during the first season after
fertilization has been established by many researchers. The median rates
of N_2O loss from fertilizer emissions vary from 0.05% for nitrate fertilizer (Burford and Hall 1977, Denmead et al. 1979, Breitenbeck et al.
1980, Conrad and Seiler 1980, Seiler and Conrad 1981, Conrad et al. 1983)
to 0.1% for ammonium salts or ammonium nitrate (Breitenbeck et al. 1980,
Conrad and Seiler 1980, Freney et al. 1981a, Seiler and Conrad 1981,
Ryden 1981, Conrad et al. 1983, McKenney et al. 1978) to 0.5% for anhydrous NH_3 or urea (Hutchinson and Mosier 1979, Breitenbeck et al. 1980,
Duxbury et al. 1982). Each of these emission factors varies by a factor
of five around the median value.
      The global usage of N fertilizer is around $5 \cdot 10^{13}$ g, which is used
on $1.4 \cdot 10^{13}$ m^2 of agricultural land (Hauck 1981). This fertilizer con-
sists of perhaps 30% urea and 20% ammonia and ammonium salts with the
remainder unidentified (Hauck 1983, FAO 1983). Thus, the N_2O flux from
fertilizer on agricultural soils is around 0.01 g $N/(m^2 \cdot yr)$. A regional calculation for Asia, assuming that 80% of the fertilizer used is
urea, would give the same N_2O emission rate from agricultural lands of
0.01 g $N/(m^2 \cdot yr)$.

2.8. RATES OF NO_x AND N_2O PRODUCTION BY LIGHTNING

      Noxon (1976) has found enhanced column densities of tropospheric
NO_2 in association with cloud-to-ground lightning over Colorado. The

fixation efficiency is estimated at approximately 10^{26} molec NO_2 per
lightning stroke. Other studies of nitrogen fixation by lightning are
based on laboratory measurements and theoretical models. Two factors are
important—the NO_x produced per unit of energy dissipated and the total
energy dissipated in lightning. These factors have been carefully reviewed by Logan (1983) and Levine et al. (1983).

The NO_x produced per unit of energy dissipated seems to be between 2
and 17 · 10^{16} molec/joule (Chameides et al. 1977, Chameides 1979, Levine
et al. 1981, Peyrous and Lapeyre 1982). Estimates of the total energy
dissipated by lightning vary from 10^{-9} to 10^{-7} joules/(cm^2 · sec) (Chameides et al. 1977, Tuck 1976, Chameides 1979, Hill et al. 1980, Levine
et al. 1981, Peyrous and Lapeyre 1982, Dawson 1980). Thus, the NO_x
production rate of lightning may vary by as much as 3 orders of magnitude. However, Logan (1983) infers that the most probable value is 8 Tg
N/yr, with a range of 2 to 20 Tg N/yr.

Turman and Edgar (1982) have presented satellite measurements of the
global distribution of lightning. Although there seem to be some incon-
sistencies in their paper between their Table 1 and the mean global
lightning-flash rate, the data clearly indicate the location of NO_x
production by lightning. The lightning flash per unit area appears to be
negligible at latitudes higher than 60° N and 60° S. Also the mean
lightning-flash rate per unit is about 9 times larger over continental
areas than over oceans. Thus, using the global NO_x production rates
quoted earlier would give NO_x production rates by lightning for continental areas between latitudes 60° S and 60° N of 5 (1 to 12) · 10^{-2} g
N/(m^2 · yr) and for remote oceanic areas, 6 (2 to 15) · 10^{-3} g N/(m^2 ·
yr).

Hill et al. (1984) have estimated the global N_2O production by
lightning and the N_2O production over North America from corona discharge
from high-voltage power lines. Because the average production rates of
these sources are small, < 0.001 Tg/yr and 0.02 Tg/yr, respectively,
these sources are ignored in the subsequent discussion.

2.9. EMISSIONS OF NO_x AND N_2O FROM COMBUSTION OF VEGETATION

Perhaps the most comprehensive summary of burnt biomass available is that
of Seiler and Crutzen (1980). They give figures for tropical, temperate,
and boreal forests; savanna and bushland; and agricultural lands. I have
used their estimates and assumed that temperate woodlands, shrublands,
and grasslands burned with a frequency of burning (or equivalent annual
fractional area burnt) of 100/yr. I also assumed that the biomass burnt
in these fires was 5 · 10^3 kg of dry matter/ha.

The annual emissions within various burnt areas are listed in
Table 2-4. Individual estimates vary over a range of at least a factor
of two from these values. The estimates of fuel-N conversion efficiency
used in Table 2-4 are from the discussion in subsection 2.3. The total
NO_x emission from burning biomass is about a factor of 2 lower than that
of other recent estimates (Logan 1983, Crutzen 1983) primarily because
the estimates in this table are based solely on data from large-scale
fires.

The N_2O production from burning biomass was evaluated using the
estimates of burnt biomass (Table 2-4) and the average emission factor

Table 2-4. Annual Emissions of NO_x-N Based on Biomass Burnt, Fuel N, and Conversion Efficiency.

Area	Biomass Burnt (10^{15} g dm)	Fuel N (%)	Conversion Efficiency (%)	NO_x-N Emission (10^{12} g N)
Forests				
Tropical	0.8 – 2.0	1	13	1.0 – 2.6
Temperate	0.2 – 0.3	1	13	0.3 – 0.4
Boreal	0.0 – 0.1	1	13	0.0 – 0.1
Woodlands, shrub lands, grasslands				
Tropical	2.0 – 3.7	0.6	10	1.2 – 2.2
Temperate	0.1	0.6	10	0.1
Agricultural land	1.7 – 2.1	0.6	10	1.0 – 1.3

Source: Estimates for biomass burnt are from Seiler and Crutzen (1980); those for fuel N are from Clark and Rosswall (1981), Galbally and Freney (1982), Robertson et al. (1982), and Rosswall (1980).

presented in subsection 2.3. I will give the emission rates for various areas in the final summary of this paper. The global annual emission of N_2O-N from burning biomass is 1-2 Tg N/yr (Crutzen 1983).

2.10. NO_x FROM THE STRATOSPHERE

The loss of N_2O in the atmosphere proceeds via two reactions:

$$N_2O + h\nu \ \longrightarrow \ N_2 + O(^1D) \quad \text{and} \tag{2-6}$$

$$N_2O + O(^1D) \longrightarrow N_2 + O_2 \tag{2-7}$$

$$\longrightarrow 2NO.$$

Calculations by Johnston et al. (1979) and Levy et al. (1979) suggest that this sink for N_2O is about 10 ± 3 Tg N/yr. The error is based on the uncertainties about the solar radiation flux, the chemical reaction rates, and the atmospheric distribution of N_2O. Some stratospheric NO is converted to NO_2 and HNO_3--all three of which are ultimately returned to the troposphere. There is some other minor production of odd nitrogen in the stratosphere (Jackman et al. 1980), as well as a sink, via two reactions of atomic nitrogen. The net source of NO_x to the troposphere is estimated to be 0.5 Tg to 1 Tg N/yr (Levy et al. 1979, Logan et al. 1981). Stratospheric air is injected into the troposphere primarily between 30° N and 90° N and between 30° S and 60° S (Gidel and Shapiro 1980). There is no evidence of different injection over oceans than over land. Throughout this region the NO_x injection from the stratosphere is estimated to be $2-5 \cdot 10^{-3}$ g N/($m^2 \cdot$ yr), an important source of NO_x in the upper troposphere.

2.11. EMISSIONS OF NH_3

Over the last decade, many studies have been published on NH_3 emissions from land surfaces. However all of these are for agricultural lands and many are specifically concerned with the loss of NH_3 from NH_3, $(NH_4)_2SO_4$, and urea fertilizer.

On a very broad scale, Lenhard and Gravenhorst (1980) have measured NH_3 emissions from rural West Germany using measurements from aircraft. These measurements probably reflect the emissions from an area of 10^4 km^2. The yearly average emission is ~ 1 g N/($m^2 \cdot$ yr). This emission rate could be accounted for by the excrement from domestic animals in the region.

On a more local scale, Denmead et al. (1974, 1976) have measured a daily emission rate in grazed pastures equivalent to 5 g N/($m^2 \cdot$ yr) with 22.5 sheep/ha and 9 g N/($m^2 \cdot$ yr) with 50 sheep/ha. Using the figures given by Healey et al. (1970) for urea excretion from sheep, these data would indicate that about 26% of the nitrogen excreted in the urine of sheep escapes to the atmosphere as NH_3. From their measurements of NH_3 emissions from an ungrazed grass-clover pasture, Denmead et al. (1976) find that, although NH_3 is released to the air at the soil surface at rates comparable to grazed pastures (presumably through decomposition of old plant material), almost all of it is absorbed by the plants growing above. Net emissions to the atmosphere measured intermittently over a period of 6 months are about 1 g N/($m^2 \cdot$ yr). Some of the NH_3 absorbed by plants from the air could subsequently be released at the end of the growing season. Farquhar et al. (1979) have found that NH_3 can be released to the air by senescing plants. They report rates equivalent to the release of 0.7 mg N/($m^2 \cdot$ day) for a leaf area index of 1. Using this figure for the NH_3 yield from senescing vegetation and assuming that the senescence occurs over 30 days, I calculated an annual input to the atmosphere of approximately $2 \cdot 10^{-2}$ g N/($m^2 \cdot$ yr).

Other Australian work concerns NH_3 losses from fertilized, cropped soils. Emissions vary from 1% of the nitrogen applied (Denmead et al. 1977) for the injection of anhydrous NH_3 to 30% (V. R. Catchpoole and O. T. Denmead, unpublished data available from Dr. Catchpoole, CSIRO Div. of Trpoical Crops and Pastures, Brisbane, Australia) for the surface application of urea. Denmead et al. (1978) have shown that the net NH_3 exchange for a corn crop varies with soil-moisture conditions and with the NH_3 being released during moist conditions and taken up during dry. Assuming that any active exchange is confined to the daytime and that the crop grows for 90 days, these data suggest a potential NH_3 exchange on the order of +0.3 to -0.2 g N/($m^2 \cdot$ yr).

A study of the release of NH_3 when fertilizer is applied to a crop via irrigation water (Denmead et al. 1982) suggests that between 5% and 20% of the NH_3 applied in this way can be volatilized to the atmosphere. Other studies of the use of $(NH_4)_2SO_4$ and urea fertilizers in flooded rice fields (Freney et al. 1981b, 1983, Simpson et al. 1984) show that between 7% and 20% of the N applied via irrigation water is lost to the

atmosphere. This does not include the NH_3 lost during the manufacture,
storage, and transport of the fertilizer.

From the data presented on the use of global fertilizers, I estimated that around 2 Tg NH_3-N was volatilized from agricultural land, or
around 0.1 g N/(m^2 · yr). A regional budget for Asia where urea usage
was high would give an estimate for the NH_3 emission rate of 0.24 g
N/(m^2 · yr) caused by fertilizer use on agricultural land.

According to the available measurements, NH_3 emissions from animal
excreta appear to dominate the global release of NH_3. However, there are
no measurements of NH_3 emissions from natural ecosystems. The emission
rates derived in this paper for agricultural lands were larger than,
although similar to, the emissions from undisturbed land surfaces derived
from the simplified model of soil processes in Dawson's 1977 article.

2.12. EMISSIONS OF NO, N_2O, AND NH_3 OVER THE OCEANS

There are no direct measurements available of NO, N_2O, or NH_3 fluxes over
any oceans. However, some other approaches have been used to evaluate
these fluxes.

NO is also produced in the tropical Pacific Ocean, probably due to
NO_2^- photolysis (Zafiriou et al. 1980, Zafiriou and McFarland 1981).
Very high supersaturations of NO are produced within the ocean but the
majority of the NO produced appears to be consumed by radical recombination within the surface water. The flux of NO to the atmosphere is estimated to be around $1 · 10^{-3}$ g N/(m^2 · yr) in the equatorial Pacific.
Measurements have not yet been made at latitudes other than those around
the equator.

N_2O is produced in the ocean and many scientists have measured the
supersaturation of N_2O in ocean surface water (Craig and Gordon 1963,
Pierotti and Rasmussen 1980, Hahn 1974, Yoshinari 1976, Cohen and Gordon
1979, Elkins et al. 1978). Several estimates of the N_2O flux from oceans
use local measurements of surface-water supersaturations of N_2O and a
boundary-layer model of the ocean-atmosphere gas transfer. Because this
approach requires globally accurate representative values of the N_2O
supersaturation and the exchange rate, neither of which are well known,
the N_2O flux is equivalently uncertain.

Cohen and Gordon (1979) and Elkins et al. (1978) have used a different approach by estimating the N_2O produced in oceans from the ratio
of N_2O to NO_3^- produced during nitrification and total NO_3^- production
rate in the oceans. As a reasonable first approximation, they suggest
that all this N_2O is released to the atmosphere and that the total amount
ranges from 4 Tg to 10 Tg N/yr. This corresponds to a global average
flux of 1 to $3 · 10^{-3}$ g N/(m^2 · yr), which is appropriate to the open
ocean.

Recently Seitzinger et al. (1983) and Smith and De Laune (1983)
have reported on their research on N_2O production in lakes and coastal
zones. Because of the intensive biological activity in these regions,
they may be a significant global source of N_2O. Fluxes of 0.01 to 0.03 g
N/(m^2 · yr) are found in these regions, or around ten times more than for
the average flux from the open ocean.

The principles regulating the flux of NH_3 between ocean and atmosphere are well understood. The factors involved include the solubility

coefficient of NH_3, its equilibrium coefficient with NH_4^+ in aqueous solution, the pH, and the temperature. Junge (1957) discusses the possibility that the ocean around Hawaii may be a source of NH_3. More recently Georgii and Gravenhorst (1977) and Ayers and Gras (1980) have continued these discussions. However, probably because of the lack of good measurement techniques, no one has yet presented a systematic set of measurements of NH_3 in surface air over the oceans along with the necessary measurements to evaluate the equilibrium partial pressure of NH_3 in ocean surface water. Until this is done, we cannot start to evaluate whether the oceans are sources or sinks of gaseous NH_3.

2.13. CONCLUSIONS

The data on emissions of NO_x, N_2O, and NH_3 into the atmosphere in various areas are presented in Table 2-5. In preparing this table I selected only those data determined by direct measurement (e.g., N_2O fluxes over land) or substantial inference (e.g., NO_x production by lightning). Where the value is missing in Table 2-5, data are not available and measurements need to be made to determine the emission.

Because only some scattered measurements of NO_x from soils are available, more measurements of these fluxes are needed from different areas, particularly in the tropics. Further measurements of NO_x

Table 2-5. Fixed Nitrogen Emissions in Remote Areas.

Area	Emissions (g N/[m² · yr])				
	N_2O-N	Soil NO_x-N	Lightning NO_x-N	Stratosphere NO_x-N	Combustion NO_x-N
Forests					
Tropical	0.2	*	0.05	0	0.07–0.17
Temperate	0.007–0.09	0.004	0.05	0.002–0.005	0.02
Boreal	*	*	0	0.002–0.005	0–0.02
Woodlands, shrub, lands, grasslands					
Tropical	0.01–0.1	*	0.05	0	0.06–0.11
Temperate	0–0.1	-0.006–0.08	0.05	0.002–0.005	0.003
Agricultural land	0.02–3.0	0.04–0.26	0.05	0.002–0.005	0.06–0.08
Tundra and alpine meadows	*	*	0	0.002–0.005	0
Swamps and meadows	0.01–0.05	*	0.05	0.002–0.005	0
Desert	*	*	0.05	0	0

Note: The only available value for NH_3-N emissions from soil and plants, including animal excrement, is 0.4–1.0 g N/(m² · yr) for agricultural land.

*Not available.

production in lightning are also needed as well as of NO_x produced by large-scale fires, particularly as related to the fuel-N content. Although good preliminary information is available on the fluxes of N_2O, further work is needed to find out why N_2O fluxes vary significantly from site to site (e.g., Duxbury et al. 1982) and to determine, on an area average basis, the relative contributions of the extreme values of N_2O release versus those contributions from values closer to the median rate.

There is an appalling lack of NH_3 emission measurements for all areas except agricultural lands. Current knowledge of the atmospheric cycle of NH_3 is based on supposition and measurements of the deposition of $NH_3 + NH_4^+$ found in rain samples. Measurements of NH_3 emissions in natural ecosystems are urgently needed.

Several sources of atmospheric NO_x, N_2O, and NH_3 have been omitted from this paper (e.g., fossil-fuel combustion)—some because little is known about them, others because they are outside our definition of remote areas. The emissions NO_x, N_2O, and NH_3 from urban areas are mainly the result of combustion processes. Crutzen (1983) and Logan (1983) discuss these emissions. In the background atmosphere, the most outstanding flux omitted was that of NH_3 from burning vegetation. Given that N is inevitably present in a reduced form (protein) within plant material, a significant fraction of the N must be volatilized as NH_3 during the heating and burning of plant material. Unfortunately no observations of this release are yet available.

2.14. REFERENCES

Allison, F. E. 1965. Evaluation of incoming and outgoing processes that affect soil nitrogen. In Soil Nitrogen (W. V. Bartholomew and F. E. Clark, eds.), Madison, WI:American Society of Agronomy, 573-606.

Anderson, R. V., D. C. Coleman, and C. V. Cole. 1981. Effects of saprotrophic grazing on net mineralization. In Terrestrial Nitrogen Cycles: Processes, Ecosystem Strategies and Management Impacts (F. E. Clark and T. Rosswall, eds.), Ecol. Bull: Stockholm 33:201-215.

Ayers, G. P., and J. L. Gras. 1980. Ammonia gas concentration over the Southern Ocean. Nature 284:539-540.

Bentley, B. L. (Chairman) et al. 1982. Report of the work group on Latin American forests. Plant and Soil 67:415-420.

Blackmer, A. M., and J. M. Bremner. 1976. Potential of soil as a sink for atmospheric nitrous oxide. Geophys. Res. Lett. 3:739-742.

Boto, K. G., and J. S. Bunt. 1982. Carbon export from mangroves. In The Cycling of Carbon, Nitrogen, Sulfur and Phosphorus in Terrestrial and Aquatic Ecosystems (I. E. Galbally and J. R. Freney, eds.), Canberra: Australia Academy of Science, 105-110.

Braun Wilke, R. H. 1982. Net primary productivity and nitrogen and carbon distribution in two xerophytic communities in central-west Argentina. Plant and Soil 67:315-324.

Breitenbeck, G. A., A. M. Blackmer, and J. M. Bremner. 1980. Effects of different nitrogen fertilizers on emission of nitrous oxide from soil. Geophys. Res. Lett. 7:85-88.

Bremner, J. M., and A. M. Blackmer. 1981. Terrestrial nitrification as a source of atmospheric nitrous oxide. In Denitrification, Nitrification and Atmospheric Nitrous Oxide (C. C. Delwiche, ed.), New York:Wiley, 151–170.

Bremner, J. M., and S. G. Robbins, and A. M. Blackmer. 1980. Seasonal variability in emission of nitrous oxide from soil. Geophys. Res. Lett. 7:641–644.

Broadbent, F. E., and F. Clark. 1965. Denitrification. In Soil Nitrogen (W. V. Bartholomew and F. E. Clark, eds.), Madison, Wisc: American Society of Agronomy, 344–359.

Bryan, B. A. 1981. Physiology and biochemistry of denitrification. In Soil Nitrogen (W. V. Bartholomew and F. E. Clark, eds.), Madison, Wisc: American Society of Agronomy, 67–84.

Bulla, L. A., C. M. Gilmour, and W. B. Bollen. 1970. Nonbiological reduction of nitrite in soil. Nature 225:664.

Burford, J. R., and K. C. Hall. 1977. Fluxes of Nitrous Oxide from a Ryegrass Sward. Agricultural Research Council, Letcombe Lab. 1976 Annual Rept. (Australia), 85–88.

Burrows, W. H. 1972. Productivity of an arid zone shrub (Eremophila gilesii) community in south-western Queensland. Aust. J. Bot. 20:317–329.

Campbell, I. M. 1977. Energy and the Atmosphere. London:Wiley, 398 pp.

Chalk, P. M., and C. J. Smith, 1983. Chemodenitrification. Development and Soil Sciences 9:65–90.

Chameides, W. L. 1979. Effect of variable energy input on nitrogen fixation in instantaneous linear discharges, Nature 277:123–125.

Chameides, W. L., D. H. Stedman, R. R. Dickerson, D. W. Rusch, R. J. Cicerone. 1977. NO_x production in lightning. J. Atmos. Sci. 34:143–149.

Cicerone, R. J., J. D. Shetter, D. H. Stedman, T. J. Kelly, and S. C. Liu. 1978. Atmospheric N_2O: Measurements to determine its sources, sinks and variations. J. Geophys. Res. 83:3042–3050.

Clark, F. E., and T. Rosswall (eds.). 1981. Terrestrial Nitrogen Cycles: Processes, Ecosystem Strategies and Management Impacts. Stockholm: Ecol. Bull., No. 33, 714 pp.

Clements, H B., and C. K. McMahon. 1980. Nitrogen oxides from burning forest fuels examined by thermogravimetry and evolved gas analysis. Thermochim. Acta 35:133–139.

Cohen, Y., and L. I. Gordon. 1979. Nitrous oxide production in the ocean. J. Geophys. Res. 84:347–353.

Conrad, R., and W. Seiler. 1980. Field measurements of the loss of fertilizer nitrogen into the atmosphere as nitrous oxide. Atmos. Environ. 14:555–558.

Conrad, R., W. Seiler, and G. Bunse. 1983. Factors influencing the loss of fertilizer nitrogen into the atmosphere as N_2O. J. Geophys. Res. 88:6709–6718.

Craig, H., and L. I. Gordon. 1963. Nitrous oxide in the ocean and marine atmosphere. Geochim. Cosmochim. Acta 27:949–955.

Crutzen, P. J. 1983. Atmospheric interactions – homogeneous gas reactions of C, N and S containing compounds. In Scope 24: The Major Biogeochemical Cycles and Their Interactions (B. Bolin and R. B. Cook, eds.) New York:Wiley Interscience, 67–111.

Crutzen, P J., L. E. Heidt, J. P. Krasnec, W. H. Pollock, and W. Seiler.
    1979. Biomass burning as a source of atmospheric gases CO, H_2, N_2O,
    NO, CH_3Cl and COS. Nature (London) 282:253–256.
Crutzen, P. J., A. C. Delany, J. Greenberg, P. Haagenson, L. Heidt, R.
    Lueb, W. Pollock, W. Seiler, A. Wartburg, and P. Zimmerman. 1985.
    Tropospheric chemical composition measurements in Brazil during the
    dry season. J. Atmos. Chem. 2:233–256..
Dawson, G. A. 1977. Atmospheric ammonia from undisturbed land. J. Geophys.
    Res. 82:3125–3133.
Dawson, G. A. 1980. Nitrogen fixation by lightning. J. Atmos. Sci.
    37:174–178.
Delany, A. C., and A. F. Wartburg. 1983. Large scale biomass burning as a
    source of oxides of nitrogen to the tropical continental troposphere.
    Paper presented at the CACGP Symposium on Tropospheric Chemistry, 28
    August to 3 September 1983, Oxford, England.
Denmead, O T., J. R. Simpson, and J R. Freney. 1974. Ammonia flux into
    the atmosphere from a grazed pasture. Science 185:609–610.
Denmead, O. T., J. R. Freney, and J. R. Simpson. 1976. A closed ammonia
    cycle within a plant canopy. Soil Biol. Biochem 8:161–164.
Denmead, O. T., J. R. Simpson, and J R. Freney. 1977. A direct field
    measurement of ammonia emission after injection of anhydrous ammonia,
    Soil. Sci. Soc. Am. J. 41:1001–1004.
Denmead, O. T., R. Nulsen, and G. W. Thurtell. 1978. Ammonia exchange
    over a corn crop. Soil Sci. Soc. Am. J. 42:840–842.
Denmead, O. T., J. R. Freney, and J. R. Simpson. 1979. Studies of nitrous
    oxide emission from a grass sward. Soil Sci. Soc. America J. 43:726–
    728.
Denmead, O T., J R. Freney, and J. R. Simpson. 1982. Dynamics of ammonia
    volatilization during furrow irrigation of maize. Soil Sci. Soc. Am.
    J. 46:149–155.
Duxbury, J M., D. R. Bouldin, R. E. Terry, and R. L. Tate, III. 1982.
    Emissions of nitrous oxide from soils. Nature 298:462–464.
Elkins, J. W., S. C. Wofsy, M. B. McElroy, C. E. Kolb, and W. A. Kaplan.
    1978. Aquatic sources and sinks for nitrous oxide. Nature 275:602–
    606.
Evans, L. F., I. A. Weeks, A. J. Eccleston, and D. R. Packham. 1977.
    Photochemical ozone in smoke from prescribed burning of forests.
    Environ. Sci. Tech. 11:896–900.
FAO, 1983. FAO Fertilizer Yearbook: 1982. FAO Statistics Series No. 48,
    Rome: UN Food and Agricultural Organization, 143 p.
Farquhar, G. D., R. Wetselaar, and P. M. Firth. 1979. Ammonia volatiliza-
    tion from senescing leaves of maize. Science 203:1257–1258.
Farquhar, G. D., P. M. Firth, R. Wetselaar, and B. Weir. 1980. On the
    gaseous exchange of ammonia between leaves and the environment:
    Determination of the ammonia compensation point. Plant Physiol.
    66:710–714.
Farquhar, G. D., R. Wetselaar, and B. Weir. 1983. Gaseous nitrogen
    losses from plants. Developments in Plant and Soil Sciences 9:159–
    180.
Fillery, I. R. P. 1983. Biological denitrification. Developments in
    Plant and Soil Sciences 9:33–64.

Floate, M. J. S. 1981. Effects of grazing by large herbivores on nitrogen cycling in agricultural ecosystems. In Terrestrial Nitrogen Cycles Process: Ecosystem Strategies and Management Impacts (F. E. Clark and T. Rosswall, eds.), Stockholm:Ecol. Bull. 33:585–602.

Franzke, C. J., and A. N. Hume. 1945. Liberation of HCN in sorghum. J. Am. Soc. Agron. 37:848–851.

Freney, J. R., O. T. Denmead, and J. R. Simpson. 1978. Soil as a source or sink for atmospheric nitrous oxide. Nature (London) 273:530–532.

Freney, J. R., J. R. Simpson, and O. T. Denmead. 1981a. Ammonia volatilization. In Terrestrial Nitrogen Cycles Process: Ecosystem Strategies and Management Impacts (F. E. Clark and T. Rosswall, eds.), Stockholm: Ecol. Bull. 33:291–302.

Freney, J. R., O. T. Denmead, I. Watanabe, and E. T. Craswell. 1981b. Ammonia and nitrous oxide losses following applications of ammonium sulphate to flooded rice. Austral. J. Agric. Res. 32:37–45.

Freney, J. R., J. R. Simpson, and O. T. Denmead. 1983. Volatilization of ammonia. Developments in Plant and Soil Sci. 9:1–32.

Galbally, I. E., and J. R. Freney (eds.). 1982. The Cycling of Carbon, Nitrogen, Sylfur and Phosphorous in Terrestrial and Aquatic Ecosystems. Canberra: Australian Academy of Science, 153 p.

Galbally, I. E., and C. R. Roy. 1978. Loss of fixed nitrogen from soils by nitric oxide exhalation. Nature (London) 275:734–735.

Galbally, I. E., and C. R. Roy. 1981. Ozone and nitrogen oxides in the southern hemisphere troposphere. In Procs. Quadrennial International Ozone Symp. (J. London, ed.), International Ozone Commission, 431–438.

Galbally, I. E., and C. R. Roy. 1983a. The fate of nitrogen compounds in the atmosphere. Developments in Plant and Soil Sci. 9:265–284.

Galbally, I. E., and C. R. Roy. 1983b. Nitric oxide exhalation from a pasture soil. Presented at the CACGP Symposium on Tropospheric Chemistry, 28 August to 3 September 1983, Oxford, England (paper available from author).

Georgii, H. W., and G. Gravenhorst. 1977. The ocean as source or sink of reactive trace-gases. Pure Appl. Geophys. 115:503–511.

Gidel, L. T., and M. O. Shapiro. 1980. General circulation model estimates of the net vertical flux of ozone in the lower stratosphere and the implications for the tropospheric ozone budget. J. Geophys. Res. 85:4049–4058.

Hahn, J. 1974. The North Atlantic Ocean as a source of atmospheric N_2O. Tellus 26:160–168.

Hatch, A. B. 1955. The influence of plant litter on the jarrah forest soils of the Dwellingup region – Western Australia. Commonw. Aust. For. Timb. Bur. Lflt. No. 70, Canberra, 18 p.

Hauck, R. D. 1981. Nitrogen fertilizer effects on nitrogen cycle processes. In Terrestrial Nitrogen Cycles Process: Ecosystem Strategies and Management Impacts (F. E. Clark and T. Rosswall, eds.), Stockholm:Ecol. Bull. 33:551–561.

Hauck, R. D. 1983. Agronomic and technological approaches to minimizing gaseous nitrogen losses from croplands. Developments in Plant and Soil Sci. 9: 285–312.

Healey, T. V., H. A. C. McKay, A. Pilbeam, and D. Scargill. 1970. Ammonia and ammonium sulfate in the troposphere over the United Kingdom. J. Geophys. Res. 75:2317-2321.

Hill, R. D., R. G. Rinker, and H. D. Wilson. 1980. Atmospheric nitrogen fixation by lightning. J. Atmos. Sci. 37:179-192.

Hill, R. D., R. G. Rinker, and A. Concouvinos. 1984. Nitrous oxide production by lightning. J. Geophys. Res. 89:1411-1421.

Hingston, F. J., and R. J. Raison. 1982. Consequences of biogeochemical interactions for adjustment to changing land use practices in forest systems. In The Cycling of Carbon, Nitrogen, Sulfur, and Phosphorus in Terrestrial and Aquatic Ecosystems (I. E. Galbally and J. R. Freney, eds), Canberra: Australian Academy of Science, 11-24

Hooper, A. B. 1978. Nitrogen oxidation and electron transport in ammonia-oxidizing bacteria. In Microbiology (D. Schlessinger, ed.), Washington, D.C.:American Society for Microbiology, 299-304.

Hooper, A. B., and K. R. Terry. 1979. Hydroxylamine oxidoreductase of nitrosomonas production of nitric oxide from hydroxylamine. Biochim. Biophys. Acta 571, 12-20.

Hummel, J., and R. Reck. 1979. A global surface albedo model. J. Appl. Meteor. 18:239-253.

Hutchinson, G. L., and A. R. Mosier. 1979. Nitrous oxide emissions from an irrigated cornfield. Science 205:1125-1127.

Ingraham, J. L. 1981. Microbiology and genetics of denitrifiers. In Denitrification, Nitrification and Atmospheric Nitrous Oxide (C. C. Delwiche, ed.), New York:Wiley, 45-66.

Jackman, C. H., J. E. Frederick, and R. S. Stolarski. 1980. Production of odd nitrogen in the stratosphere and mesophere: An intercomparison of source strengths. J. Geophys. Res. 85:7495-7505.

Johansson, C. 1984. Field measurements of emission of nitric oxide from fertilized and unfertilized forest soils in Sweden, J. Atmos. Chem. 1:429-442.

Johansson, C., and I. E. Galbally. 1985. The production of nitric oxide in a loam under aerobic and anaerobic conditions, Appl. and Environ. Microbial. (in press).

Johansson, C., and L. Granat. 1984. Emission of nitric oxide from arable land. Tellus 36B:25-37.

Johnston, H. S., O. Serang, and J. Podolske. 1979. Instantaneous global nitrous oxide photochemical rates. J. Geophys. Res. 84:5077-5082.

Judeikis, H. S., and A. G. Wren. 1978. Laboratory measurements of NO and NO_2 depositions onto soil and cement surfaces. Atmos. Env. 12:2315-2319.

Junge, C. E. 1957. Chemical analysis of aerosol particles and of gas trace on the Island of Hawaii. Tellus 9:528-537.

Keller, M., T. J. Goreau, S. C. Wolfsy, W. A. Kaplan, and M. B. McElroy. 1983. Geophys. Res. Lett. 10:1156-1159.

Knowles, R. 1981. Denitrification. In Terrestrial Nitrogen Cycles, Process, Ecosystem Strategies and Management Impacts (F. E. Clark and T. Rosswall, eds.), Stockholm:Ecol. Bull. 33:315-330. Lenhard, U., and G. Gravenhorst. 1980. Evaluation of ammonia fluxes into the free atmosphere over western Germany. Tellus 32:48-55.

Lenhard, U., and G. Gravenhorst. 1980. Evaluation of ammonia fluxes into the free atmosphere over western Germany. Tellus 32:48-55.

Levine, J. S., R. S. Rogowski, G. L. Gregory, W. E. Howell, and J. Fishman. 1981. Simultaneous measurements of NO_x, NO and O production in
a laboratory discharge: Atmospheric implications. Geophys. Res.
Letts. 8:357-360.

Levine, J. S., T. R. Augustsson, I. C. Anderson and J. M. Hoell, Jr.,
1983. The NO_x budget of troposphere. Paper presented at the CACGP
Symposium on Tropospheric Chemistry, August 28 to September 3,
Oxford, England.

Levy, I. I., J. D. Mahlman, and W. J. Moxim. 1979. A preliminary report on
the numerical simulation of the three-dimensional structure and
variability of atmospheric N_2O. Geophys. Res. Letts. 6:155-158.

Lipschultz, F., O. C. Zafiriou, S. C. Wofsy, M. B. McElroy, F. W. Valois,
and S. W. Watson. 1981. Production of NO and N_2O by soil nitrifying
bacteria. Nature 294:641-643.

Logan, J. A. 1983. Nitrogen oxides in the troposphere: Global and regional budgets. J. Geophys. Res. 88:10,785-10,807.

Logan, J. A., M. J. Prather, S. C. Wofsy, and M. B. McElroy, 1981. Tropospheric chemistry: A global perspective. J. Geophys. Res. 86:7210-
7254.

Manskaya, S. M., and T. V. Drozdova. 1968. Geochemistry of organic substances. Oxford:Pergamon Press, 345 p.

Matthews, E. 1983. Global vegetation and land use: New high-resolution
data bases for climate studies, J. Climate Appl. Meteorol. 22:474-
487.

McKenney, D. J., D. L. Wade, and W. I. Findlay. 1978. Rates of N_2O evolution from N fertilized soil. Geophys. Res. Lett. 5:777-780.

McKenney, D. J., K. F. Shuttleworth, and W. I. Findlay. 1980. Temperature
dependence of nitrous oxide production from Brookston Clay. Can. J.
Soil. Sci. 60:665-674.

McKenney, D. J., K. F. Shuttleworth, J. R. Vriesacker, and W. I. Findlay.
1982. Production and loss of nitric oxide from denitrification in
anaerobic Brookston Clay. Appl. Environ. Microbiol. 43:534-541.

Medina, E. 1982. Nitrogen balance in the Trachypogan grasslands of central Venezuela. Plant and Soil 67:305-314.

Moore, A. W., J. S. Russell, and J. E. Coaldrake. 1967. Dry matter and
nutrient content of a subtropical semiarid forest of Acacia harpo-
phylla F. Muell (Brigalow). Aust. J. Bot. 15:11-24.

Moraghan, J. T., and R. J. Buresh. 1977. Chemical reduction of nitrite
and nitrous oxide by ferrous iron. Soil Sci. Soc. Amer. J. 41:47-50.

Mulcahy, M. F. R. 1973. Gas Kinetics. Surrey, England:Nelson, 305 pp.

Nelson, D. W., and J. M. Bremner. 1970. Gaseous products of nitrite
decomposition in soils. Soil Biol. Biochem. 2:203-215.

Nommik, N., and J. Thorin, 1972. A mass spectrometric technique for
studying the nitrogeneous gases produced on the reaction of nitrite
with raw humus. Agrochimica 16:319-322.

Noxon, J. F. 1976. Atmospheric nitrogen fixation by lightning. Geophys.
Res. Letts. 3:463-465.

Olson, J., and J. A. Watts. 1982. Major World Ecosystem Complexes. Oak
Ridge, TN:Oak Ridge National Laboratory.

Payne, W. J. 1981. The status of nitric oxide and nitrous oxide as inter-
      mediates in denitrification. In **Denitrification, Nitrification and
      Atmospheric Nitrous Oxide** (C. C. Delwiche, ed.), New York:Wiley, 85–
      103.
Peyrous, R., and R.-M Lapeyre. 1982. Gaseous products created by elec-
      trical discharges in the atmosphere and condensation nuclei resulting
      from gaseous phase reactions. **Atmos. Environ** 16:959–968.
Pierotti, D., and R. A. Rasmussen. 1980. Nitrous oxide measurements in
      the eastern tropical Pacific Ocean. **Tellus** 32:56–72.
Ritchie, G. A. F., and D. J. D. Nicholas. 1972. Identification of the
      sources of nitrous oxide produced by oxidative and reductive pro-
      cesses in **Nitrosomonas** europaea. **Biochim. J.** 126:1181–1191.
Robertson, G. P., R. Herrera, and T. Rosswall (eds.). 1982. **Nitrogen
      Cycling in Ecosystems of Latin American and the Caribbean: Develop-
      ments in Plant and Soil Sciences**, The Hague: Martinus Nijhoff/Dr. W.
      Junk Publishers, Vol. 6, 430 p.
Rogers, H. H., J. C. Campbell, and R. J. Volk. 1979. Nitrogen-15 dioxide
      uptake and incorporation by phaseolus vulgaris (L.). **Science** 206:333–
      335.
Rosswall, T. 1980 (ed.). **Nitrogen Cycling in West African Ecosystems.**
      SCOPE/UNEP International Nitrogen Unit, Stockholm: Royal Swedish
      Academy of Sciences, 450 p.
Rosswall, T. 1982. Microbiological regulation of the biogeochemical ni-
      trogen cycle. **Plant Soil** 67:15–34.
Roy, C. R. 1979. Atmospheric nitrous oxide in the mid-latitudes of the
      southern hemisphere. **J. Geophys. Res.** 84:3711–3718.
Ryden, J. C. 1981. N_2O exchange between a grassland soil and the atmo-
      sphere. **Nature** 292:235–237.
Seiler, W., and R. Conrad. 1981. Field measurements of natural and fer-
      tilizer-induced N_2O release rates from soils. **J. Air Poll. Control
      Assoc.** 31:767–772.
Seiler, W., and P. J. Crutzen. 1980. Estimates of the gross and net
      fluxes of carbon between the biosphere and the atmosphere from bio-
      mass burning. **Climatic Change** 2:207–247.
Seitzinger, S. P., M. E. Q. Pilson, and S. W. Nixon. 1983. Nitrous oxide
      production in nearshore marine sediments, **Science** 222:1244–1246.
Simpson, J R., J. R. Freney, R. Wetselaar, W. A. Muirhead, R. Leuning, and
      O. T. Denmead. 1984. Transformations and losses of urea nitrogen
      after application to flooded rice. **Aust. J. Agric. Res.** 35:189–200.
Skujins, J. 1981. Nitrogen cycling in arid ecosystems. In **Terrestrial
      Nitrogen Cycles. Process. Ecosystem Strategies and Management Impacts**
      (F. E. Clark and T. Rosswall, eds.), Stockholm:Ecol. Bull. 33:477–
      492.
Slemr, F., and W. Seiler, 1983. Field measurements of NO and NO_2 emission
      from soil. Paper presented at the CACGP Symposium on Tropospheric
      Chemistry, 28 August to 3 September 1983, Oxford, England.
Smith, C. J., and R. D. De Laune. 1983. Nitrogen loss from freshwater and
      saline estuarine sediments. **J. Environ. Qual.** 12:514–518.

Smith, C. J., R. D. De Laune, and W. H. Patrick. 1982. Carbon and nitrogen cycling in a _spartina alterniflora_ salt marsh. In _The Cycling of Carbon, Nitrogen, Sulfur, and Phosphorus in Terrestrial and Aquatic Ecosystems_ (I. E. Galbally and J. R. Freney, eds), Canberra: Australian Academy of Science, 97–104.

Steen, W. C., and B. J. Stojanovic. 1971. Nitric oxide volatilization from a calcareous soil and model aqueous solutions. _Soil Sci. Soc. Amer. Proc._ 35:277–282.

Tuck, A. F. 1976. Production of nitrogen oxides by lightning discharges. _Quart. J. Roy. Met. Soc._ 102:749–755.

Turman, B. N., and B. C. Edgar. 1982. Global lightning distributions at dawn and dusk. _J. Geophys. Res._ 87:1191–1206.

Van Cleemput, O., W. H. Patrick, Jr., and R. C. McIlhenny. 1976. Nitrite decomposition in flooded soil under different pH and redox potential conditions. _Soil Sci. Soc. Amer. J._ 40:55–60.

Van Cleve, K., and V. Alexander. 1981. Nitrogen cycling in tundra and boreal ecosystems. In _Terrestrial Nitrogen Cycles, Process, Ecosystem Strategies and Management Impacts_ (F. E. Clark and T. Rosswall, eds.), Stockholm:Ecol. Bull. 33:375–404.

Verstraete, W. 1981. Nitrification. In _Terrestrial Nitrogen Cycles, Process, Ecosystem Strategies and Management Impacts_ (F. E. Clark and T. Rosswall, eds.), Stockholm:Ecol. Bull. 33:303–314.

Vines, R. G., L. Gibson, A. B. Hatch, N. K. King, D. A. MacArthur, D. R. Packham, and R. J. Taylor. 1971. _On the Nature, Properties and Behavior of Bushfire Smoke._ CSIRO Division of Applied Chemistry Tech. Paper No. 1, Mordialloc, Australia:CSIRO, 32 p.

Wetselaar, R. 1980. Nitrogen cycling in a semi-arid region of tropical Australia. In _Nitrogen Cycling in West African Ecosystems_ (T. Rosswall, ed.), SCOPE/UNEP International Nitrogen Unit, Stockholm: Royal Academy of Sciences, 157–170.

Wetselaar, R., and G. D. Farquhar. 1980. Losses of nitrogen from the tops of plants. _Adv. Agron._ 23:263–302.

Wetselaar, R., and J. T. Hutton. 1963. The ionic composition of rain water at Katherine, N.T., and its part in the cycling of plant nutrients. _Aust. J. Agric. Res._ 14:319–329.

Yates, L. R. (Chairman) et al. 1982. Report on the work group on savannas and shrublands. _Plant and Soil_ 67:409–413.

Yoshinari, T. 1976. Nitrous oxide in the sea. _Marine Chemistry_ 4:189–202.

Zafiriou, O. C., M. McFarland, and R. H. Bromund. 1980. Nitrous oxide in seawater. _Science_ 207:637–639.

Zafiriou, O. C., and M. McFarland. 1981. Nitric oxide from nitrite photolysis in the central equatorial Pacific. _J. Geophys. Res._ 86:3173–3182.

3. THE EMISSION OF SULFUR AND NITROGEN TO THE REMOTE ATMOSPHERE WORKING-GROUP REPORT

Ian E. Galbally
CSIRO, Division of Atmospheric Research
Private Bag No. 1
Mordialloc, Victoria
3195 Australia

Meinrat O. Andreae
Department of Oceanography
Florida State University
Tallahassee, FL 32306

Bernard Bonsang
Centre des Faibles Radioactivités
Gif-sur-Yvette
91190 France

Paul J. Crutzen, Director
Airchemistry Department
Max-Planck-Institut für Chemie
P. O. Box 3060
D-6500 Mainz
Federal Republic of Germany

Sherry O. Farwell
Department of Chemistry
University of Idaho
Moscow, ID 83843

Elvira Tsani
General Chemical State Laboratory
Division of Environmental Pollution Control
Athens 11521
Greece

3.1. INTRODUCTION

Sulfur and nitrogen compounds are emitted to the remote atmosphere from many varied biological and geological sources in the form of their diverse chemical species. Since the atmospheric fate of the emitted compounds depends on their chemical reactivities with other molecules in the atmosphere, we considered the emissions of sulfur and nitrogen in

55

J. N. Galloway et al. (eds.), The Biogeochemical Cycling of Sulfur and Nitrogen in the Remote Atmosphere, 55–63.
© *1985 by D. Reidel Publishing Company.*

terms of chemical species rather than in terms of elements. To represent
accurately the temporal and spatial variabilities of the various sources,
they must be understood in terms of processes rather than as ''black
boxes'' so that emission measurements can be placed into a biogeographical and ecological framework.

In the discussions at the meetings of the working group on emissions
in Bermuda, we first tried to establish a conceptual and methodological
framework that would define the requirements placed upon the accuracy of
emission estimates by the fate of the individual chemical species in the
global cycles and the techniques by which these requirements could be
met. We then discussed how much was known about the emissions from
different sources of the key species of sulfur and nitrogen. In these
discussions we also attempted as much to identify the most significant
gaps in the available information as to summarize the state-of-the-art.

3.2. CONCEPTUAL AND METHODOLOGICAL FRAMEWORKS

3.2.1. Accuracy Requirements for Emission Estimates

Our working group considered the emissions of dimethylsulfide (DMS), CS_2,
COS, H_2S, SO_2, sulfate, NO_x, N_2O, and NH_3 to the atmosphere. For regional cycling studies, measurements of the emissions of most of these
species should be precise to $\pm 5 \cdot 10^{-3}$ g (S or N)/($m^2 \cdot$ yr) (signal/
noise = 1) with an accuracy of $\pm 10\%$. The exceptions to this were COS,
CS_2, and N_2O because the longer atmospheric lifetimes (combined with the
relevant atmospheric burdens) involved require emission measurements to
be precise to $\pm 5 \cdot 10^{-4}$ g (S or N)/($m^2 \cdot$ yr) to estimate adequately
source terms. (CS_2 was included since it is a precursor to the long-
lived COS.) These precisions were selected so that the uncertainty in
global emission rates for each species from imprecise measurements would
only be approximately 1% of the global S or N flux, i.e., ~ 1.0 Tg and
0.1 Tg (S or N)/yr of the global flux for short- and long-lived species,
respectively.

There are very few measurements of the emission rates of the species in which we were interested. For many, no measurements have been
made over many of the world's continents or oceans. We agreed that the
ideal data requirement for acceptable global studies would be at least
one flux measurement taken over a year's time that would be representative of the area of each major ecosystem and of each biogeographic region
of the oceans of the world.

The available data on S and N emissions from soils show that the
areal variability is large with variations in emission rates greater than
a factor of 10 over distances of a few meters (Galbally and Roy 1983, see
also Chapters 1 and 2). Other information on N emissions suggests that
there are also large temporal variations of N emissions (Johansson and
Granat 1984, see also Chapter 2). Therefore, measurement programs
designed to provide regional estimates of emissions of any of these
species are needed that incorporate adequate spatial and temporal sampling. Sampling strategies based on a classification system, such as an
ecosystem- or soil-classification model, are also needed.

Ignorance pervaded our understanding of NH_3 emissions from all
natural systems. A reliable, specific, continuous method is needed that

would measure NH_3 at sub-ppb levels before the emissions and concentrations of NH_3 in remote areas can be determined. We did not even know the molecular forms of all N and S species emitted to the atmosphere—e.g., CH_3SCHCH_2 (methyl vinyl sulfide) might be emitted from soil (unpublished data available from Dr. S. O. Farwell, University of Idaho, Moscow). Also there are no reliable measurements of the emissions of other known species, including MeSH (methylmercaptan), HCN, CH_3CN, and amines.

Most S and N compounds emitted to the atmosphere are, through physical and biological processes, highly influenced by local meteorological conditions, e.g., rain and temperature. As our knowledge increases, it may be possible to include such coupling into our understanding and modeling of atmospheric composition.

3.2.2. Analytical Techniques

We agreed that the well-established procedures of analytical chemistry need to be used in all verifications or validations of analyses of any constituent under consideration (ACS 1983). Particular attention should be paid to the elimination or reduction of interferences. Two or more independent measurement techniques, collaborative measurements, and quality-assurance programs should be used and a full error analysis should be part of any data presentation. These issues have been inadequately addressed in the past in this field and such protocols must be an essential part of all present and future measurement programs.

3.2.3. Flux Measurements

Techniques for obtaining emission rates over soil/plant and water surfaces include (1) static and flow-through enclosures (chambers), (2) micrometeorological-flux/profile and eddy-correlation measurements, (3) mass-balance measurements near a discontinuous source, and (4) concentration-gradient/transfer-velocity calculations for air/water interfaces. Each technique has its own advantages and limitations. Adequate protocols are urgently needed for each technique and for the intercomparison of simultaneous measurements from the different techniques.

The measurement of emission rates from forests, particularly tropical forests, presents particular problems. The flux/profile method has known deficiencies that preclude its use over forests with irregular canopies (Stewart and Thom 1973). Similar reservations were discussed about the accuracy of eddy-correlation measurements taken above forests. The enclosure technique is not readily usable in a forest with a dense high canopy nor does it give representative results for a tropical ecosystem because of the wide variety of trees.

The only suitable technique for measuring emissions from tropical forests appeared to be the mass-balance method that uses aircraft measurements under conditions of steady air-flow with a capping inversion. However, for this technique to be applicable, the lifetime of the species being measured must be known because of chemical reactions in the ambient air and deposition. If the chemical-transformation time constant is

comparable to the time for air to pass over the forest (fetch divided by
wind speed), then even this technique may be unusable.

3.2.4. Gross-versus-Net Fluxes

In some current regional and global cycle studies, a conceptual model is
used in which there are simultaneous emission and dry-deposition processes occurring over the same region (see Chapters 6 and 8). Within the
model, each process is an irreversible, one-way movement of material.
The use of these gross fluxes may exaggerate the magnitude of the atmospheric circulation of the element in question and, therefore, the use of
a net flux might be more appropriate.

   A two-way flow is found where (1) the source and sink mechanisms
within one medium are identical (e.g., the NH_3 release from and uptake by
ocean water) and (2) the source and sink mechanisms within one medium are
different (e.g., the microbial release of NO on land but the major uptake
by plants). In the latter case, gross fluxes do have some ecological
significance whereas, in the former, they are meaningless.

   Because of the bidirectional character of exchange fluxes for such
gases, it seemed inappropriate to use the concept of an undirectional
deposition velocity for NH_3 over land and oceans and for NO_x over land to
estimate source and sink fluxes. It would be much more appropriate to
use the net fluxes based either on direct measurements or, over oceans,
on measurements of NH_3 in both oceans and the atmosphere and then to use
transfer-velocity calculations.

3.3. EMISSIONS OF SULFUR COMPOUNDS

3.3.1. Sulfur Emissions to the Marine Atmosphere

We discussed sulfur fluxes to the marine atmosphere and agreed that DMS
emission is a source of sulfur to the remote marine atmosphere with
emissions of 40 (20 to 60) Tg S/yr (Andreae and Raemdonck 1983 and
Chapter 1). The marine DMS source has high spatial variability (by about
a factor of 10 from lowest to highest), which is related to the spatial
variability of certain phytoplankton species (Barnard et al. 1984).

   We estimated that the deposition of excess non-sea-salt sulfate to
the ocean is 55 (40–70) Tg S/yr (see Chapters 8 and 9), which suggests
that DMS is the major source of sulfur in the marine atmosphere. This
does not mean that other S species are not also making a significant
contribution. Although H_2S fluxes from the ocean has not yet been measured, H_2S is possibly being emitted from surface layers. Also the
fluxes of DMDS and other reduced species have not yet been quantified but
the available data suggest that the DMDS flux is very small (unpublished
data available from Dr. M. O. Andreae, Florida State University,
Tallahassee).

   COS is produced in seawater by photochemically initiated reactions
with dissolved organic matter that presumably contains S (Ferek and
Andreae 1984). This source of COS, when combined with other sources
(those from volcanoes, oxidation of CS_2, production of COS in mineral
sulfide ores, biomass burning, and emission from soils), exceeds the

stratospheric photolysis sink for COS (Crutzen et al. 1979). Uptake by plants is the most probable additional sink (Taylor et al. 1983). Measurements of the major ion composition of sea-salt aerosol indicate a significant enrichment by non-sea-salt sulfate (Andreae 1982). However, we still do not know the source of this enrichment or the processes that are causing it.

3.3.2. Sulfur Emissions to the Remote Continental Atmosphere

Sulfur gas fluxes from soil-plant systems have been measured at dryland, freshwater-wetland, and saline-wetland sites. However, we had serious reservations about whether meaningful areal averages could be composed based on the present data base. The measurements of H_2S and MeSH emissions from the chamber experiments by Adams et al. (1981), the largest such study, might be too low because of moisture condensation in the chamber (personal communication from Dr. S. O. Farwell, University of Idaho, Moscow). Since H_2S at the ppt level can be reliably measured using cryogenic trapping with gas-chromatographic analyses and $AgNO_3$ trapping with fluorescence analyses (Herrmann and Jaeschke 1984), measurement technology is not a limiting factor. We also need reliable measurements of H_2S emissions over the continents, including wetlands and the tropics.

There is much to be learned about such fluxes, including whether littoral macrophytes (porous plants) are playing a role in transferring reduced sulfur compounds from anoxic soil zones to the atmosphere. Preliminary reports on the few systems that have been investigated suggest that such plant species are constituting the predominant ventilatory route for sediment gases, such as CH_4, CS_2, and COS (Aneja et al. 1979, Dacey and Klug 1979).

3.3.3. Volcanic Emissions of Sulfur

Although large volcanic eruptions are very spectacular and release large amounts of S to the atmosphere, emissions from volcanoes in the noneruptive phase might be making the major contribution to emission fluxes. Berresheim and Jaeschke (1983) calculate an average global emission of 12 Tg (S)/yr from volcanoes. This is a long-term average of what could be a highly fluctuating time series since, in years of major eruptions, the emission might increase substantially. Two points were made in our discussions about the estimate by Berresheim and Jaeschke (1983): (1) The older data used in their calculations are based on inadequate sampling and measurement techniques, which would cause emission rates to be underestimated, and (2) their only available record of volcanic eruptions is incomplete, which again would contribute to an underestimation of emission rates. Therefore, we felt that the global release rate of 12 Tg S/yr is probably a lower limit for the volcanic flux.

We agreed that indications of the volcanic S contribution need to be looked for in concentrations and depositions at selected remote sites downwind of active but noneruptive volcanoes. Because of its enrichment in volcanic volatiles (Gerlach and Gräber 1985), fluorine might be useful as a tracer of volcanic emissions.

3.4. EMISSIONS OF NITROGEN COMPOUNDS

3.4.1. Nitrogen Oxides and Ammonia Emissions to the Marine Atmosphere

There is considerable uncertainty about the ocean/atmosphere flux of N_2O
(see also Chapter 2). The degree of supersaturation or undersaturation of
N_2O in ocean surface waters has been extensively measured by Weiss
(unpublished data available from Dr. R. Weiss, Scripps Institute of
Oceanography, La Jolla, CA). Weiss's work should lead to more reliable
estimates of the ocean-to-atmosphere flux of N_2O if calculations from his
data and the concentration-gradient/transfer-velocity approach are used.
     Sources of NO_x over the ocean include NO_2^- photolysis within the
ocean, lightning, stratospheric/tropospheric exchange, and long-range
transport from continents (Chapter 2). The data on aerosol nitrate found
in samples from Fanning Island in the equatorial Pacific and of rainwater
from Amsterdam Island show evidence of the long-range transport of NO_x
(Savoie 1984, Galloway and Gaudry 1984).
     Unpublished calculations (available from Dr. D. Kley, Institut für
Chemie der Kernforschungsanlge, Jülich, FRG) based on new data suggest
that at least half the NO_x produced by lightning is in the upper tropical
troposphere and readily available for long-range transport. This production is confined between latitudes 30° S to 30° N. Within this region,
the production rates are comparable to those already published (Logan
1983 and Chapter 2).
     Comparison of sources over and deposition from remote tropical oce-
anic areas for NO_x (including HNO_3 and NO_3 aerosol) suggested that an
additional source of NO_x might be contributing to the N cycle in remote
oceanic areas (Table 3-1, see also Chapters 8 and 10). This is consistent with the suggestions concerning the long-range transport of NO_x
(including HNO_3, PAN, and other nitrates) from continental areas.
     There are insufficient good measurements of NH_4^+ concentrations in
upper layers of open ocean water and NH_3 in the gas phase above an ocean
to calculate directly the air/ocean exchange of NH_3. However, the
NH_4HSO_4 found in marine aerosol (Andreae 1982, see also Chapter 1) and
NH_4^+ in precipitation (see Chapter 8) indicate the presence of such an
exchange. Nevertheless, our ignorance of the origin and rate-limiting
processes for the supply of this NH_3 to the remote marine atmosphere is
disconcerting. Careful measurements need to be made of NH_4^+ and NH_3 in
ocean surface water along with simultaneous measurements of NH_3 in the
oceanic atmosphere so that the ocean/atmosphere fluxes of NH_3 may be
quantitatively determined.

3.4.2. Nitrogen Emissions to the Remote Continental Atmosphere

There is some information on nitrogen oxide emissions in remote continental regions in the tropics. This information is summarized in Table
2-5 (Chapter 2). There may be an excess of NO_x emissions over depositions in remote continental regions and this excess may be transported to
the marine atmosphere. Experiments are needed to measure the transport

Table 3-1. Sources and Sinks of NO_x, including HNO_3 and aerosol NO_3, in Remote Oceanic and Continental Areas.

	NO_x (g N/[m^2 · yr])	
	Oceanic	Continental
Sources		
Lightning	0.002–0.015	0.01–0.12
Stratosphere	0.002–0.005	----
Seawater	0.001 (?)	----
Biomass burning	----	0.05–0.2
Soil release	----	?
Totals	0.005–0.02.	0.06–0.32 (+?)
Sinks		
Wet deposition	0.01–0.10	0.03–0.13
Dry deposition	0.02–0.03	0.01–0.04
Totals	0.03–0.13	0.04–0.17

of NO_x over remote continental boundaries to detect and quantify this flux.

3.5. SUMMARY AND RECOMMENDATIONS

Some major sources of atmospheric sulfur, including volcanoes and marine emissions of DMS, have been identified and quantified to an uncertainty of less than an order of magnitude. However, we knew so little about other sulfur sources over the remote continents and the oceans that, at the time of the Bermuda workshop, we could not say whether the source and removal fluxes for atmospheric sulfur are reasonably well balanced.

There were even greater uncertainties about the emissions of nitrogen species. We knew nothing about the direct emission rates of NH_3 from either natural ecosystems or oceans. There is some information on emission rates of N_2O from natural ecosystems but far less is known about NO_x.

The methodology exists for measuring most of these emission rates but there is a dearth of field measurements. Because so much more is known about the removal of S and N from the atmosphere, emission measurements are urgently needed to verify or contradict our understanding of these fluxes.

We specifically recommend that

- Comprehensive protocols be developed for both constituent analyses and flux measurements.

- Different flux-measurement techniques be intercompared.

- Ecological characteristics be established for the environments where fluxes are being measured.

- Areal variabilities of emissions **within** **and** **between** ecosystems be determined.

- Emissions of all relevant S and N species be measured in tropical ecosystems and tropical ocean regions.

- Specific additional emphasis be placed on obtaining NH_3 emission measurements.

- Field studies of NO_x production in lightning be conducted.

- Total emission rates of S and N be established so that the presence or absence of emissions from presently unstudied species may be determined.

3.6. REFERENCES

ACS. 1983. Principles of environmental analysis. Anal. Chem. 55:2210–2218.

Adams, D. F., S. O. Farwell, E. Robinson, M. R. Pack, and W. L. Bamesberger. 1981. Biogenic sulfur source strengths. Environ. Sci. Technol. 15:1493–1498.

Andreae, M. O. 1982. Marine aerosol chemistry at Cape Grim, Tasmania, and Townsville, Queensland. J. Geophys. Res. 87:8875–8885.

Andreae, M. O., and H. Raemdonck. 1983. Dimethylsulfide in the surface ocean and the marine atmosphere: A global view. Science 221:744–747.

Aneja, V.P., J. H. Overton, L. T. Cupitt, J. L. Durham, and W. E. Wilson. 1979. Direct measurement of emission rates of some atmospheric biogenic sulfur compounds. Tellus 31:174–178.

Barnard, W. R., M. O. Andreae, and R. L. Iverson. 1984. Dimethylsulfide and Phaeocystis poucheti in the southeastern Bering Sea. Cont. Shelf Res. 3:103–113.

Berresheim, H., and W. Jaeschke. 1983. The contribution of volcanoes to the global atmospheric sulfur budget. J. Geophys. Res. 88:3732–3740.

Crutzen, P. J., L. E. Heidt, J. P. Krasnec, W. H. Pollock, and W. Seiler. 1979. Biomass burning as a source of atmospheric gases CO, H_2, N_2O, NO, CH_3Cl and COS. Nature (London) 282:253–256.

Dacey, J. W. H., and M. J. Klug. 1979. Efflux from lake sediments through waterlilies. Science 203:1253–1254.

Ferek, R. J., and M. O. Andreae. 1984. Photochemical production of
    carbonyl sulfide in marine surface waters. Nature 307:148–150.
Galbally, I. E., and C. R. Roy. 1983. Nitric oxide exhalation from a
    pasture soil. Paper presented at CACGP Symposium on Tropospheric
    Chemistry (abstracts only published), 28 August to 3 September 1983,
    Oxford, England.
Galloway, J. N., and A. Gaudry. 1984. The composition of precipitation on
    Amsterdam Island, Indian Ocean. Atmos. Environ 18:2649–2656.
Gerlach, T. M., and E. J. Gräber. 1985. Volatile budget of Kilauea vol-
    cano. Nature 313:173–277.
Herrmann, J., and W. Jaeschke. 1984. Measurements of H_2S and SO_2 over
    the Atlantic Ocean. J. Atmos. Chem. 1:111–123.
Johansson, C., and L. Granat. 1984. Emission of nitric oxide from arable
    land. Tellus 36B:25–37.
Logan, J. 1983. Nitrogen oxides in the troposphere: Global and regional
    budgets. J. Geophys. Res. 88:10785–10807.
Savoie, D. L. 1984. Nitrate and non–sea–salt sulfate aerosols over major
    regions of the world ocean: Concentrations, sources, and fluxes. Ph.
    D. diss., University of Miami, Coral Gables, FL.
Stewart, J. B., and A. S. Thom. 1973. Energy budgets in pine forest.
    Quart. J. Roy. Met. Soc. 99:154–170.
Taylor, G. S., M. B. Baker, and R. J. Charlson. 1983. Heterogeneous
    interactions of the C, N, and S cycles in the atmosphere: The role of
    aerosols and clouds. In SCOPE 24:The Major Biogeochemical Cycles
    and Their Interactions (B. Bolin and R. B. Cook, eds.), New York:
    Wiley, 115–142.

PART II

THE TRANSFORMATIONS OF SULFUR AND NITROGEN IN THE REMOTE ATMOSPHERE

4. THE TRANSFORMATIONS OF SULFUR AND NITROGEN IN THE REMOTE ATMOSPHERE BACKGROUND PAPER

Robert J. Charlson
Department of Civil Engineering
Environmental Engineering and Science Program
University of Washington
Seattle, WA 98195

William L. Chameides
School of Geophysical Sciences
Georgia Institute of Technology
Atlanta, GA 30332

Dieter Kley
National Oceanic and Atmospheric Administration*
Boulder, CO 80307

4.1. INTRODUCTION

Many, if not most, sulfur and nitrogen compounds are introduced into the troposphere as gases (see Chapters 1 and 2) and a portion of these compounds are removed directly to the earth's surface. Before their removal, most of the remaining fraction undergo both chemical transformation to another molecular form and physical transformation to a condensed phase, either in aerosol particles or in cloud droplets. Simultaneously, all the substances are mixed by eddy motions into the planetary boundary layer and, in some cases, throughout the troposphere and into the stratosphere. Volcanoes sporadically inject relatively small but important amounts of some species directly into the stratosphere where transformations also occur before slow transport back to the earth's surface via the troposphere. Because three physical states are involved (gas, solid or liquid aerosol particles, and solute in cloud droplets), the transformations occurring in each phase and between the three phases need to be described.

The gas phase of each key molecular species is clearly defined and needs no explanation. The concentration of each species alone, for example, can be used to calculate gas-phase reaction rates or vapor-pressure relationships. Both the aerosol and cloud phases, however, require more information than concentration alone to define adequately both the physical and chemical properties that control transformations.

---

*Present address: Institut für Chemie der Kernforschungsanlage Jülich GmbH, Institut 2, Postfach 1913, D-1570 Jülich, Federal Republic of Germany.

J. N. Galloway et al. (eds.), The Biogeochemical Cycling of Sulfur and Nitrogen in the Remote Atmosphere, 67–80.

4.1.1. Aerosol Morphology

Figure 4-1 schematically presents the volume distribution of tropospheric
aerosols as it is controlled by source factors, physical transformations,
and removal. This general form of the size distribution is based on
substantial data from continental air and fragmentary data from maritime,

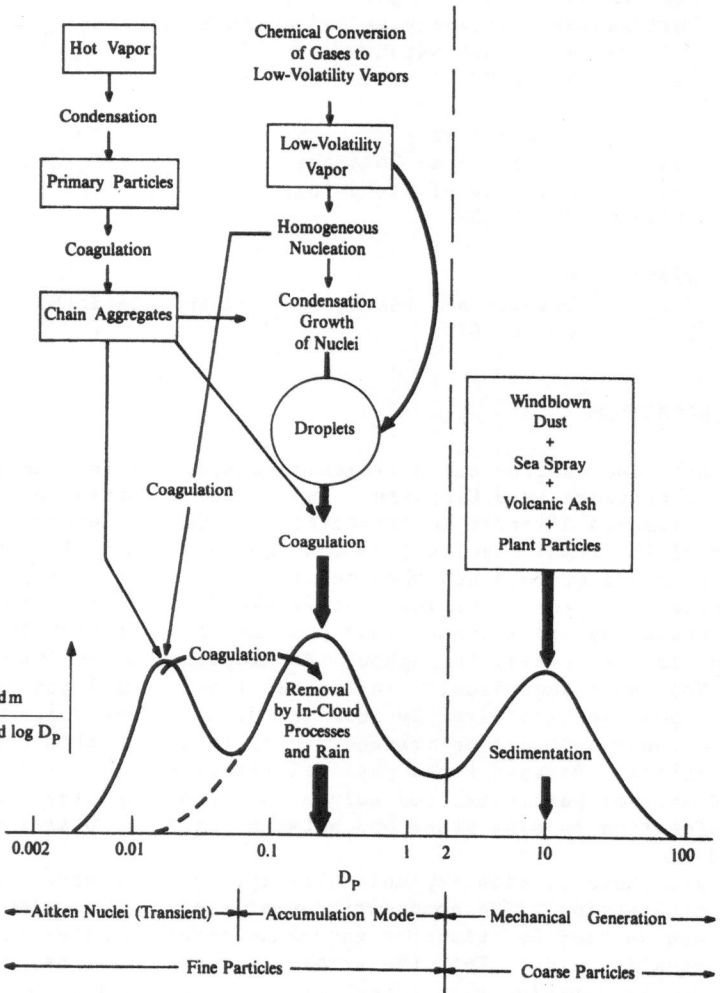

Figure 4-1.   Schematic of an Atmospheric-aerosol Size Distribution. This
     shows the three mass modes, the main sources of mass for each mode,
     and the principal processes involved in inserting mass into and
     removing mass from each mode (m = mass concentration, D_p = particle
     diameter). Data from Whitby and Sverdrup (1980).

background sites. Figure 4-2 gives an example of a volume distribution from a remote area over the Pacific Ocean.

In general, a fine-particle-aerosol mass-concentration mode exists in which the particles have diameters smaller than 2 μm (perhaps smaller than 0.5 μm in remote marine areas). This mode is, to a large degree, physically separate from the coarse-aerosol-particle mass mode and is not in chemical contact with the coarser mode. The ratio of 20 or more in particle diameter for the peaks of the two mass modes results in a ratio of individual particle masses of about 10^4 or more, which imposes radically different physical behaviors on fine and coarse particles. The different production mechanisms for fine and coarse particles indicate different production locations and, of course, different production rates. Because the fine and coarse modes vary in size and composition, each has different atmospheric processes and pathways. Because of the variety of chemical and physical transformations involved, we have used particle size as the criterion for dividing this discussion into subsections.

4.1.1.1. <u>Coarse Particles</u>. The coarse mode usually contains some $SO_4^=$ and NO_3^- and, sometimes, NH_4^+. Sulfate from sea salt is ubiquitous in the marine boundary layer. Some sulfate may also arise over land as windblown dust from exposed gypsum deposits and may, possibly, come from other sources. The observed enrichment of Ca^{++} and $SO_4^=$ in remote marine air may be partially caused by interactions of $CaCO_3$-rich particles and SO_2 (unpublished data available from Dr. M. O. Andreae, Florida State

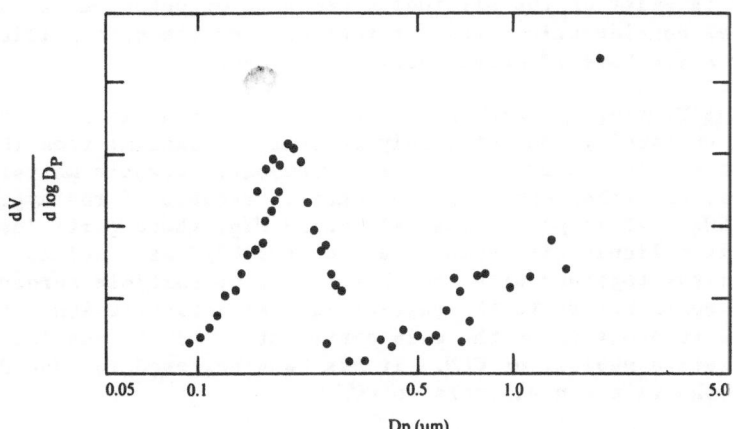

Figure 4-2. Typical Particle Volume Distribution in Marine Air Over the South Pacific (V = volume concentration). Data for particles with diameters larger than approximately 3 μm are truncated. Note that the minimum of bimodal distribution is at approximately 0.3 μm rather than at 1 μm as found in continental smog. Samples were taken aboard the Soviet vessel Korolev, 24 November 1983, latitude 10° S, longitude 135° W. (A. D. Clarke and R. J. Charlson, unpublished data available from either author, University of Washington, Seattle).

University, Tallahassee). These contribute somewhat to the atmospheric
flux of $SO_4^=$ in dry or wet deposition. Values for $SO_4^=$ from sea salt can
be derived from measurements of Na^+ or Cl^- and (assuming a purely oceanic
source) the known ratio of $SO_4^=$ to Cl^- or Na^+ in seawater. Sulfate found
in excess of the seawater ratio is called ''excess sulfate'' and is
assumed to be caused by other processes in the sulfur cycle. To ascertain a value for $SO_4^=$ emitted from land-based mineral deposits such as
gypsum is much more difficult because of the lack of a fixed reference
species such as the Na^+ and Cl^- in seawater. Therefore, caution must be
exercised when calculating excess sulfate because of the possibility of a
nonmarine source as well as measurement uncertainties.

In contrast, NO_3^- in coarse particles seems to be caused by the
interaction of $HNO_3(g)$ with the mechanically produced coarse particles.
NO_3^- tends to be found mainly in particles with diameters of between
1.0 μm and 5 μm, somewhat smaller than at the peak of the coarse-particle
mass mode. This could be the result of the reaction of HNO_3 gas with the
surface of basic sea salt or soil dust because the surface-area distribution for the coarse mode is at its peak when the particles are smaller
than they are at the peak of the mass distribution. Although there is
substantial area in the fine mode for it to act as a sink, the usual
strong acidity of the fine particles would indicate that the basic coarse
mode is the chief particulate sink for HNO_3.

Sedimentation is a major sink for coarse particles because their
source is primarily at the earth's surface and because particles with
diameters larger than a few micrometers have finite fall speeds. Fall
speed also limits the altitude to which coarse particles may rise. Although few data exist on the altitude dependence of coarse-mode species,
the mechanical considerations suggest that most coarse-mode particles may
rise to only a few hundred meters above the ground.

4.1.1.2. **Fine Particles.** Fine-particle aerosols with diameters smaller
than about 1 μm usually consist mainly of sulfates ranging from $(NH_4)_2SO_4$
to H_2SO_4 with smaller amounts of nitrate, condensed organic matter, elemental carbon, and other combustion products. Because of the hygroscopic
nature of H_2SO_4 that is partly neutralized by NH_3, these particles often
contain H_2O as a liquid with some or all of the $SO_4^=$ as a solute. This
water solubility, together with the fact that fine-particle aerosols are
the dominant contributors to the particle-number concentration, causes
fine-particle aerosols to be the main contributors to the population of
cloud condensation nuclei, or CCN. It has been believed for decades that
sulfates are the main contributors to CCN.

4.1.2. Clouds

Cloud droplets form on some of these solution droplets and aerosol particles when the relative humidity exceeds 100%. When the cloud droplets
grow through condensation, the solute is diluted by many orders of magnitude. Nonetheless, aerosol particles dissolved in cloud water are major
sources of the solute in cloud water. Subsequently, cloud droplets form
falling hydrometeors either by coalescence, vapor deposition on ice, or
by graupel (hail) formation and some of the solute is carried to the

ground by rain, hail, or snow. Typically, clouds in the lower troposphere, which contain most of the condensed water in the atmosphere,
consist mainly of liquid droplets. Typical clouds have droplet popula-
tions of approximately 10^2–10^3 droplets/cm^3, with droplet sizes ranging
from a few to a few tens of micrometers in diameter and achieving liquid-
water contents of a tenth to, at most, a few grams of water per cubic
meter. Fog may have substantially lower liquid-water content than
clouds, perhaps 0.01 g/m^3 or less. Drop-size distributions vary significantly depending on both the properties of the aerosol involved and the
meteorological conditions that produce the supersaturated air, as seen in
Figure 4-3.

Thus, the droplet-surface areas that are available for gas exchange
also vary substantially from one situation to another. Ice particles in
clouds can grow as a result of vapor deposition, and they do so at the
expense of supercooled droplets, at the same temperature, that have a
higher equilibrium water-vapor pressure. Because of growth by vapor
deposition, ice particles are both chemically different and more dilute
than droplets. Because of the relative importance of the ice phase in
precipitation formation and because both coalesced droplets and melted
snowflakes produce rain, hydrometeor composition and geochemical fluxes
vary substantially, even if the chemical composition in the atmosphere
remains constant (Scott 1981). Graupel (hail) formation does not result
in such radical composition differences. Since glaciers are formed at
least partially by snowflakes, solute concentrations found in samples
from glacial meltwater cannot be easily related to those from rainwater

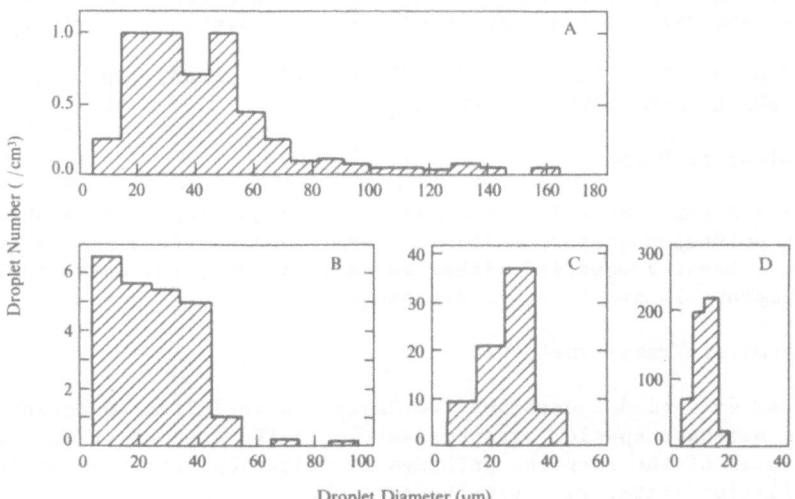

Figure 4-3. Droplet Spectra in Various Cloud Types (^W_L = liquid-water
content). Note change in ordinate scales from A-D. Samples are
from A an orographic cloud over Hawaii (^W_L = 0.40 g/m^3), B a dark
stratus over Hilo, Hawaii (^W_L = 0.34 g/m^3), C a trade-wind cumulus
over the Pacific off the Hawaiian coast (^W_L = 0.50 g/m^3), D a continental cumulus over the Blue Mountains near Sidney, Australia (^W_L =
0.35 g/m^3). Data from Pruppacher and Klett (1980).

in warmer locations because of the differences in the mechanisms of the
hydrometeor growth. This is particularly important when considering the
historical studies of ice cores from Greenland and Antarctica.

4.1.3. Interactions of Gases with Aerosol Particles and Cloud Droplets

Throughout their existence in the atmosphere, gaseous aerosol precursors
(SO_2, NO, NH_3 etc.) interact physically and chemically with both the
aerosol particles and cloud droplets (if the relative humidity exceeds
100%). The main physical interaction may consist of dissolution into the
aqueous phase where many chemical interactions are possible. The oxidation of SO_2 in the gas phase results in the formation of low-vapor-
pressure species, such as H_2SO_4, which either form new particles or
condense on the available surfaces, mainly on aerosol particles smaller
than about 0.1 μm diameter. Although water-soluble gases may react
within the aqueous phase of aerosol particles, they are more likely to
react in clouds because of the presence of so much more liquid water.

The sequence of physical and chemical transformations from gaseous
precursors to aerosol particles to clouds and then to rain, with a continuous interaction of the condensed phases with trace gases, leads to a
picture of the atmospheric cycle of trace substances. Both chemical and
physical transformations occur simultaneously. The overall physical
transformation scheme is depicted in Figure 4-4.

Because chemical transformations can also be followed as steps in an
atmospheric cycle, we can now examine them in more detail. After listing
the key molecular species, we will introduce flow diagrams and the corresponding chemical reactions for sulfur and nitrogen, respectively.

4.2. MOLECULAR FORMS, OXIDATION STATES, PHYSICAL FORMS, AND REACTIONS OF KEY SULFUR AND NITROGEN SPECIES

4.2.1. Molecular Forms

Tables 4-1 and 4-2, for sulfur and nitrogen, respectively, describe the
variety of oxidation states, molecular forms, and physical states for the
atmosphere. Species expected either to exist in very low concentration
or to be improbable are noted in parentheses.

4.2.2. Chemical Transformations

Figures 4-5 and 4-6 describe the known chemical transformations for
sulfur and nitrogen species, respectively. In the case of sulfur, the
main character of the reaction pathways is oxidation from a lower to a
higher oxidation state, with simultaneous physical conversion from a
gaseous to an aerosol or solute form. Sulfur species are not known to be
reduced in the atmosphere although COS might react to H_2S (McElroy et al.
1980). This atmospheric cycle is in many (if not most) instances, a case
of injection into the atmosphere with an oxidation state of -2, an oxidation to +4 and then to +6, and then continuous or subsequent removal.
Alternatively, the +4 state (SO_2) is injected directly but follows a
similar path. By contrast, nitrogen cycling involves substantial amounts
of chemical reduction, particularly from photochemical reactions. The

Table 4-1. Atmospheric Sulfur Compounds

Oxidation State	Molecular/Ionic Forms		
	Gas	Aerosol	Cloud/Rain
+6	$(SO_3)(H_2SO_4)$	H_2SO_4, HSO_4^-, NH_4HSO_4, $(NH_4)_2SO_4$	$SO_4^=$
+5	--	--	--
+4	SO_2	$H_2O \cdot SO_2$, HSO_3^-, CH_3SO_3H	$H_2O \cdot SO_2$, HSO_3^-, $HOCH_2SO_3^-$, $SO_3^=$ $CH_3SO_3^-$
+3			
+2	(SO)		
+1			
0			$(CH_3)_2SO$
-1			
-2	H_2S, RSH, RSR, $RSSR$ CS_2, COS	--	--

Notes: ( ) denotes either a very low concentration or an improbable
form. $R=CH_3$ appears to be dominant in almost all remote areas.

Table 4-2. Atmospheric Nitrogen Compounds

Oxidation State	Molecular/Ionic Forms		
	Gas	Aerosol	Cloud/Rain
+5	NO_3, N_2O_5, HNO_3, $R(O)O_2NO_2$	HNO_3, NO_3^-	NO_3^-
+4	NO_2, (N_2O_4)	--	--
+3	HNO_2	HNO_2, NO_2^-	NO_2^-
+2	NO	--	--
+1	N_2O	--	--
0	N_2	--	--
-1			
-2			
-3	NH_3, RNH_2, R_2NH, R_3N	NH_4^+, RNH_3^+, etc.	NH_4^+, RNH_3^+, etc.

Notes: ( ) denotes either a very low concentration or an improbable
form. $R=CH_3$ appears to be dominant in almost all remote areas.

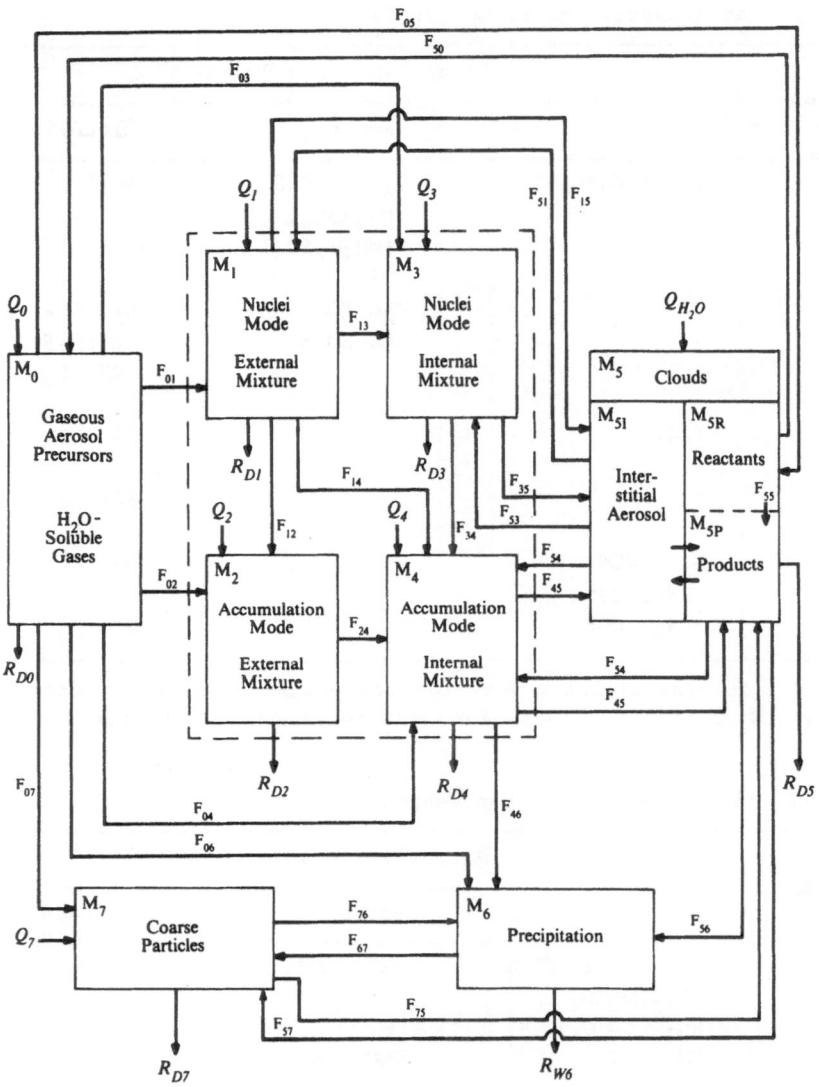

Figure 4-4.   The Physical Transformations in the Atmospheric Cycles of
     Sulfur and Nitrogen.   Each box represents a physically and chemi-
     cally definable entity in the atmosphere.   The transformations are
     given in F_{ij}.(from the ith to the jth box). Q_i represents sources
     contributing to the mass or burden, M_i, in the ith box. R_{Di} and R_{Wi}
     are dry and wet removals from M_i.   The dashed box represents what
     may be called the fine-particle aerosol and could be a single box
     instead of the set of four sub-boxes (i = 1, 2, 3, 4).   The physical
     transformations are as follows:

F_{01}: Production of new nuclei-mode particles.

F_{02}: Growth of existing accumulation-mode particles by the deposition of products of chemical reactions.

F_{03}: Growth of pre-existing nuclei-mode particles, as in F_{02}.

F_{04}: Growth of internally mixed accumulation mode, as in F_{02}.

F_{05}: Dissolution of gaseous reactants in cloud drops.

F_{50}: Reverse of F_{05}--evaporation or gaseous exchange or both.

F_{06}: Below-cloud scavenging of gaseous reactants or reaction products.

F_{07}: Interaction of gases with coarse particles, e.g., $HNO_3(g)$ + sea salt $\longrightarrow$ coarse mode NO_3^-.

F_{12}: Brownian coagulation of nuclei-mode particles with themselves to produce accumulation-size (chemically) externally mixed particles.

F_{13}: Adsorption, condensation.

F_{14}: Coagulation.

F_{15}: Cloud formation yielding nuclei-mode interstitial aerosol, coagulation with cloud droplets, or (unlikely) activation as a cloud condensation nucleus (CCN).

F_{51}: Cloud evaporation releasing interstitial aerosol.

F_{57}: Cloud evaporation releasing coarse particles.

F_{24}: Adsorption, coagulation, condensation.

F_{34}: Adsorption, coagulation, condensation.

F_{35}: Cloud formation, as in F_{15}.

F_{53}: Cloud evaporation, as in F_{51}.

F_{45}: Cloud formation, likely to be CCN.

F_{54}: Cloud evaporation releasing CCN.

F_{55}: Reactions in cloud water producing changes in solute mass.

F_{46}: Below-cloud scavenging.

F_{56}: Formation of precipitation.

F_{67}: Evaporation of precipitation particles (raindrops) before reaching ground.

F75: Coarse particles acting as CCN.

F_{76}: Below-cloud scavenging of coarse-mode particles.

R_{D5}: Occult precipitation (deposition of cloud droplets directly to the earth's surface, trees, etc.).

NH_3 cycle might also be viewed as being separate from that of NO and NO_2, particularly because of the likelihood that NH_3 will react with H_2SO_4-containing aerosol particles and be removed before being further involved in a more oxidized form of nitrogen.

In both Figures 4-5 and 4-6, the rates of reaction are omitted for simplicity. Some reactions are thought to be fast, with residence times of a species being minutes or hours, while other reactions are much slower, with residence times of days. It would be useful to have accurately known rates for all the reactions. However, using such rates in a complete model would also require knowing details of both gas-phase and condensed-phase compositions, chemical dynamics, and rates of physical transformation.

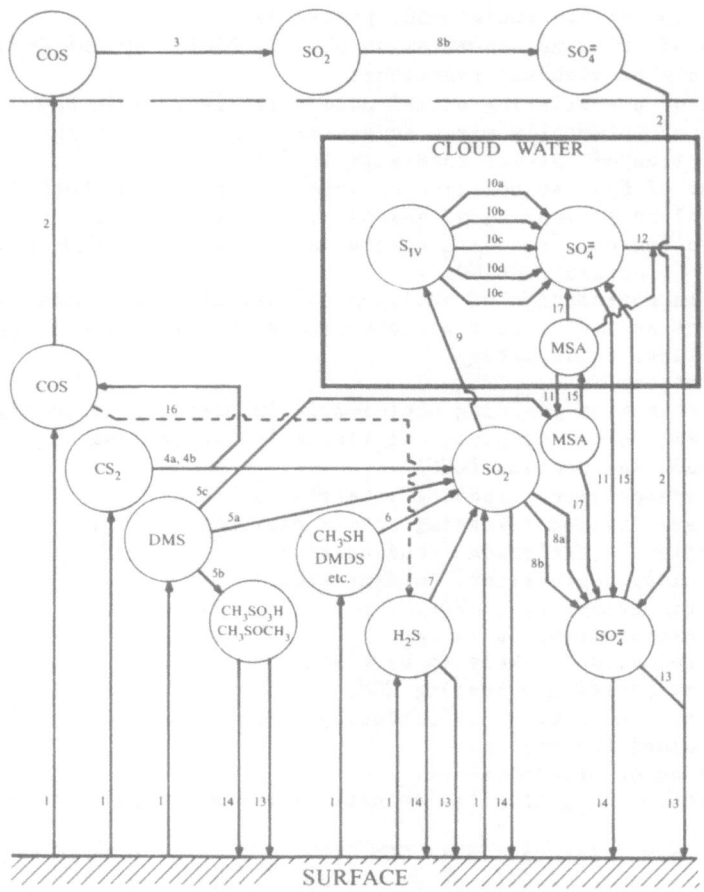

Figure 4-5. The Chemical Transformations of Sulfur in the Atmospheric
    Cycle. Circles are chemical species, the box represents cloud-liquid
    phase. DMS = CH_3SCH_3, DMDS = CH_3SSCH_3, S_{IV} = $(SO_2)_{aq}$ + HSO_3^- + $SO_3^=$ +
    + $CH_2OHSO_3^-$, and MSA (methane sulfonic acid) = CH_3SO_3H. The chemi-
    cal transformations are as follows:

1: Surface emissions
2: Tropospheric/stratospheric exchange
3: $COS + h\nu \longrightarrow S + S + CO$
   $S + O_2 \longrightarrow SO + O$
   $SO + O_2 \longrightarrow SO_2 + O$
4a: $CS_2 + OH \longrightarrow CS_2OH$
   $CS_2OH \longrightarrow$ multistep $\longrightarrow COS + SO_2$

4b: $CS_2 + h\nu \longrightarrow CS_2^*$

$CS_2^* + O_2 \longrightarrow CS + SO_2$

$CS + O_2 \longrightarrow COS + O$

$CS + O_3 \longrightarrow COS + O_2$

5a: $CH_3SCH_3 + OH \longrightarrow$ multistep $\longrightarrow SO_2$

5b: $CH_3SCH_3 + OH \longrightarrow$ multistep $\longrightarrow CH_3SOCH_3$

5c: $CH_3SCH_3 + OH \longrightarrow$ multistep $\longrightarrow CH_3SO_3H$

6: $CH_3SH + OH \longrightarrow$ multistep $\longrightarrow SO_2$

7: $H_2S + OH \longrightarrow$ multistep $\longrightarrow SO_2$

8a: $SO_2 + OH \longrightarrow HSO_3$

$HSO_3 + O_3 \longrightarrow HO_2 + SO_3$

$SO_3 + H_2O \longrightarrow H_2SO_4$

Heterogeneous catalysis
8b: $SO_2 \xrightarrow{\hspace{5cm}} SO_4^=$

9: $(SO_2)_g \longleftrightarrow (SO_2)_{aq}$

$(SO_2)_{aq} + H_2O \longrightarrow HSO_3^- + H^+$

$HSO_3^- \longleftrightarrow H^+ + SO_3^=$

$CH_2(OH)_2 + HSO_3^- \longleftrightarrow H_2O + CH_2OHSO_3^-$

10a: $(H_2O_2)g \longleftrightarrow (H_2O_2)_{aq}$

$HSO_3^- + (H_2O_2)_{aq} \longrightarrow$ multistep $\longrightarrow H^+ + SO_4^=$

10b: $(O_3)_g \longleftrightarrow (O_3)_{aq}$

$HSO_3^- + (O_3)_{aq} \longrightarrow$ multistep $\longrightarrow H^+ + SO_4^=$

10c: $(HO_2)_g \longleftrightarrow (HO_2)_{aq}$

$(HO_2)_{aq} \longrightarrow H^+ + O_2^-$

$(HO_2)_{aq} + O_2 \xrightarrow{H_2O} (H_2O_2)_{aq} + OH^-$

$HSO_3^- + (H_2O_2)_{aq} \longrightarrow$ multistep $\longrightarrow 2H^+ + SO_4^=$

10d: $(OH)_g \longleftrightarrow (OH)_{aq}$

$HSO_3^- + (OH)_{aq} \longrightarrow$ multistep $\longrightarrow 2H^+ + SO_4^=$

10e: $HSO_3^- + O_2 \longrightarrow$ multistep $\longrightarrow H^+ + SO_4^=$

11: Evaporation
12: $SO_4^=$ in cloud water $\longrightarrow SO_4^=$ in rainwater
13: Washout, rainout
14: Dry deposition
15: Cloud nucleation
16: $COS + OH \longrightarrow$ multistep $\longrightarrow H_2S$
17: MSA $\longrightarrow SO_4^=$ by some mechanism

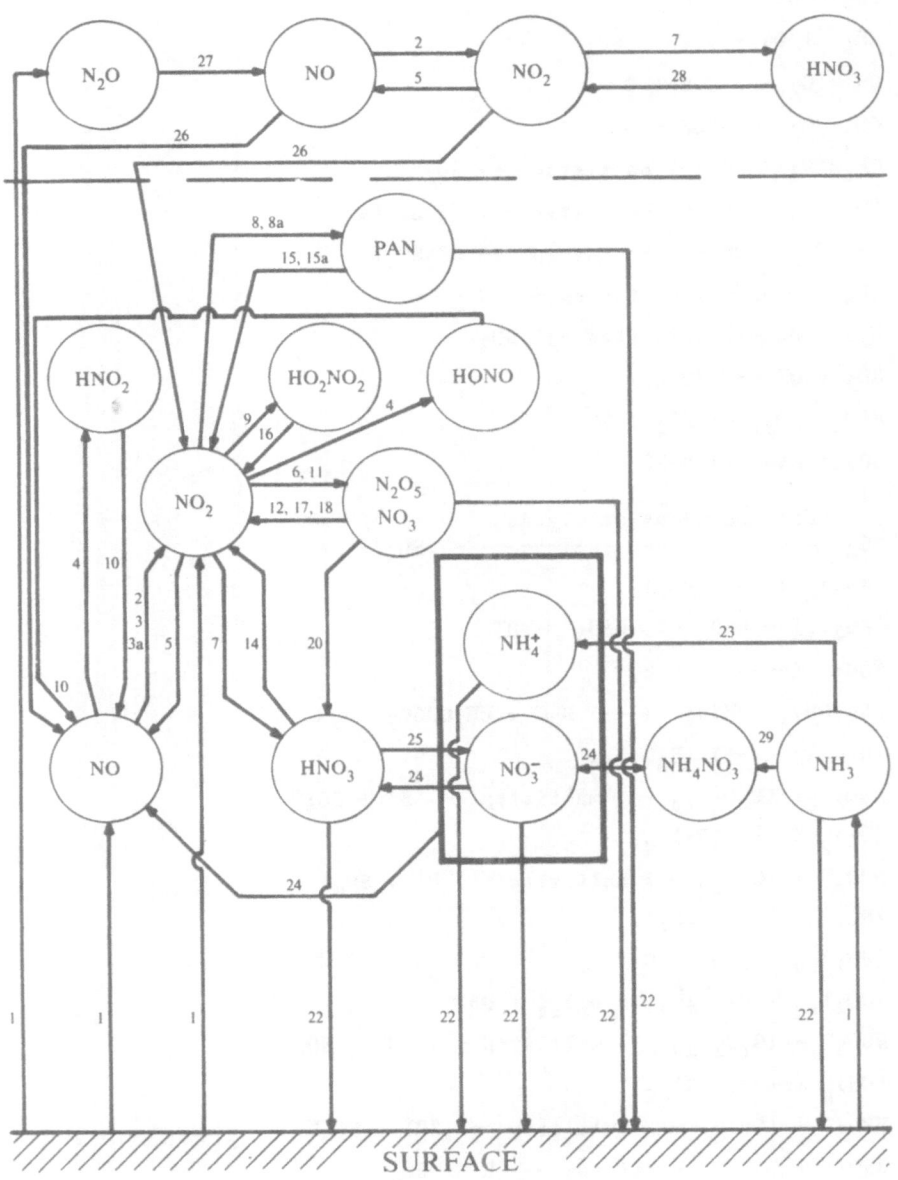

Figure 4-6.   The Chemical Transformations of Nitrogen in the Atmospheric
   Cycle.   Circles are chemical species, the box represents cloud-
   liquid phase.

1: Sources

2: $NO + O_3 \longrightarrow NO_2 + O_2$

3: $NO + HO_2 \longrightarrow NO_2 + OH$

3a: $NO + RO_2 \longrightarrow NO_2 + RO$

4: $NO + OH + M \longrightarrow HONO + M$

5: $NO_2 + h\nu \longrightarrow NO + O$

6: $NO_2 + O_3 \longrightarrow NO_3 + O_2$

7: $NO_2 + OH + M \longrightarrow HNO_3 + M$

8: $NO_2 + CH_3C(O)O_2 + M \longrightarrow CH_3C(O)O_2NO_2 + M$

8a: $NO_2 + R(O)O_2 + M \longrightarrow R(O)O_2NO_2 + M$

9: $NO_2 + HO_2 + M \longrightarrow HO_2NO_2 + M$

10: $HONO + h\nu \longrightarrow OH + NO$

11: $NO_3 + NO_2 + M \longrightarrow N_2O_5 + M$

12: $NO_3 + h\nu \longrightarrow NO_2 + O$

13: $NO_3 + x \longrightarrow products$

14: $HNO_3 + h\nu \longrightarrow NO_2 + OH$

15: $CH_3C(O)O_2NO_2 + h\nu \longrightarrow CH_3C(O)O_2 + NO_2$

15a: $R(O)O_2NO_2 \longrightarrow R(O)O_2 + NO_2$

16: $HO_2NO_2 \longrightarrow HO_2 + NO_2$

17: $N_2O_5 + h\nu \longrightarrow NO_3 + NO_2$

18: $N_2O_5 \longrightarrow NO_3 + NO_2$

19: $CH_3C(O)O_2 + NO \longrightarrow CH_3 + NO_2 + CO_2$

20: $N_2O_5 \longrightarrow HNO_3$

21: $NO_3 \longrightarrow products$

22: Sinks to surface

23: $NH_3 (g) \longleftrightarrow NH_3 (aq)$

$H_2O + NH_3 (aq) \longrightarrow NH_4^+ + OH^-$

24: Evaporation of cloud water

25: $HNO_3 (g) \longleftrightarrow HNO_3 (aq)$

$HNO_3 (aq) \longrightarrow NO_3^- + H^+$

26: Tropospheric/stratospheric exchange

27: $N_2O + O^1D \longrightarrow NO + NO$

28: $HNO_3 + h\nu \longrightarrow NO_2 + OH$

29: $NH_3 + HNO_3 \longrightarrow NH_4NO_3$

$$30: N_2 + O_2 \xrightarrow{\text{lightning}} 2NO$$

4.3.   REMAINING, CENTRAL QUESTIONS

The subject of this chapter—transformations—is indeed a large topic,
which includes many of the scientific focal points of atmospheric chemistry. In spite of this apparent complexity, it is useful to view the
atmosphere as encompassing chemical cycles of sulfur and nitrogen, with
both the chemical and physical changes that occur throughout the cycles.
By so doing, the resultant changes in molecular form are put in perspective and the relationships to sources and sinks can be assessed. The
controlling factors, such as the influence of clouds, aerosol physics,
photochemistry, free radical reactions, surface reactions, etc., are
clearly evident.
     Interestingly, the interactions of various species within either the
gaseous or condensed phase of the atmosphere can lead to important feedbacks. A simple example is the influence of NH_3 on cloud-water pH and the
resultant changes in solubility and dissociation of SO_2 in cloud water.
Feedbacks of aerosol content on liquid-water content and cloud physics
are also a possibility. Such feedbacks, along with the known or possible
second-order chemical-reaction mechanisms, suggest that the systems represented in Figures 4-4, 4-5, and 4-6 may not respond linearly. Thus,
although it might be tempting to write a simple first-order model around
the boxes in Figure 4-4, such an exercise might produce quite unrealistic
results. Therefore a next goal should be the development of a sound
understanding of chemical-reaction mechanisms, rates of physical transformations and mechanisms, and—ultimately—the extent of feedback, if
any.

4.4.   REFERENCES

McElroy, M. B., S. C. Wofry, and N. D. Sze. 1980. Photochemical sources
     for atmospheric H_2S. Atmos. Environ 14:159-163.
Pruppacher, H. R., and J. D. Klett. 1980. Microphysics of Clouds and
     Precipitation. Dordrecht:Reidel.
Scott, B. C. 1981. Sulfate washout ratios in winter storms. J. Appl.
     Meteorol. 20:619-625.
Whitby, K. T., and G. M. Sverdrup. 1980. California aerosols: Their
     physical and chemical characteristics. In The Character and Origins
     of Smog Aerosols (G. M. Hidy et al., eds.), New York:Wiley
     Interscience, pp. 499-517.

5. THE TRANSFORMATION OF SULFUR AND NITROGEN IN THE REMOTE ATMOSPHERE WORKING GROUP REPORT

Robert B. Chatfield
National Center for Atmospheric Research
P. O. Box 3000
Boulder, CO 80307

Robert J. Charlson
Department of Civil Engineering
Environmental Engineering and
   Science Program
University of Washington
Seattle, WA 98195

James P. Friend
Department of Chemistry
Drexel University
Philadelphia, PA 19104

Wolfgang Jaeschke
Center for Environmental Protection
University of Frankfurt/Main
Federal Republic of Germany

Dieter Kley
NOAA*
Boulder, CO 80307

Harold I. Schiff
Chemistry Department
York University
Downsview, Ontario
Canada M3J1P3

5.1. INTRODUCTION

In this chapter, we have tried to assess the knowledge available at the time of the Bermuda workshop on the mechanisms and rates of transformation and the ways of estimating which pathways are dominant or important. Because of the breadth of the field and the many unanswered questions at the time of this workshop, we have provided a guide rather than specific rates. We have illustrated that:

- Lifetimes of different species vary extensively and the lifetime of any one species may vary under different conditions.

- Rates of chemical reactions and physical transformations are often comparable.

*Present address: Institut für Chemie der Kernforschungsanlage Jülich GmbH, Institut 2, Postach 1913, D-1570 Jülich, Federal Republic of Germany.

J. N. Galloway et al. (eds.), The Biogeochemical Cycling of Sulfur and Nitrogen in the Remote Atmosphere, 83–101.

- For most relevant reactive species of sulfur and nitrogen, important transformations may take place over a few days thereby leading to the subglobal distribution of these species.

The transformations of chemical species involve processes that are studied in different disciplines: gas-phase radical kinetics, liquid-phase ion-radical equilibria and reactions, aerosol and cloud micro-physics, cloud dynamics, surface-layer micrometeorology (including the associated surface chemistry of absorption), and large-scale meteorology. A common error of air chemists has been to describe one sort of transfor-mation and to neglect a very different, dominating process. For example, in analyzing the progress of a sulfur atom in an SO_2 molecule from its emission near the earth's surface to its return there, the focus has often been on the fascinating complexity of its gas- and liquid-phase reactivity while ignoring the powerful effects of dry deposition to the earth's surface, which, in some cases, controls the concentration of SO_2.

This example illustrates a general point about competing processes: When a species has alternative fates, e.g.,

$$SO_2 \begin{cases} \text{Reaction in gas phase with HO} \xrightarrow{\ a\ } SO_3 \dashrightarrow H_2SO_4 \\[1em] \text{Absorption into water and} \\ \quad \text{reaction with } H_2O_2 \text{ or other oxidants} \xrightarrow{\ b\ } \text{sulfate} \\[1em] \text{Solution into rain or cloud water} \xrightarrow{\ c\ } \text{sulfite} \\[1em] \text{Absorption to the earth's surface} \xrightarrow{\ d\ } ?, \end{cases}$$

there is competition between the rates for each individual process. If we call these rates R_a, R_b, etc., and the total rate, R_{tot}, is given as the sum of the individual rates, then

$$R_{tot} = R_a + R_b + R_c + R_d \qquad (5\text{-}1)$$

would be typically dominated by one of these rate processes. Overall or partial time scales are used to describe systems like these. An overall time scale, T, may be defined as

$$T = \frac{[SO_2]}{R_{tot}} \qquad (5\text{-}2)$$

and a partial time scale, T_a, as

$$T_a = \frac{[SO_2]}{R_a} . \qquad (5\text{-}3)$$

For minor process, these individual time scales may be much longer than the overall time scale and, frequently, they have a simple form, as in gas- and liquid-phase kinetics (see Weston and Schwartz 1972). The rate of competitive processes is determined by the most rapid process, as shown above.

Consecutive processes are also important in any description of transformations. The oxidation of SO_2 by hydrogen peroxide in solution takes only seconds when the reactants are available:

$$SO_2 \cdot H_2O + H_2O_2 \xrightarrow{r} H_2SO_4. \tag{5-4}$$

However, in arriving at a reaction rate, R_b, that may be compared to the others, we must consider the consecutive processes involved for SO_2 to reach cloud liquid water and to be absorbed. We can give this process a rate, R_β, and consider its associated time scale to begin with, e.g., an SO_2 molecule moving around in a mixed layer beneath a field of cumulus clouds. If, on the other hand, the mixed layer is beneath a clear sky, this process would be altered and thus R_β. Another process would, instead, be defined--one in which such clear skies become cloudy with cumulus clouds--and the rate labelled R_α. The time scale associated with the latter process may be as much as several days and the liquid-phase oxidation process would be determined by this rate. Approximately speaking

$$T_b \simeq T_\alpha + T_\beta + T_r \simeq \max_i T_i \text{ and} \tag{5-5}$$

$$R_b \simeq \min_i R_i. \tag{5-6}$$

This example illustrates how the overall rate of a set of consecutive processes is determined by the slowest process (Weston and Schwartz 1972, see also Tables 7-1 and 7-2).

Laboratory studies have contributed a great deal to our detailed understanding of rate processes and their control by temperature, pressure, etc. Gas-phase transformation rates have been reviewed by Atkinson and Lloyd (1984) and Baulch et al. (1984) and many liquid-phase rates have been summarized by Jacob and Hoffman (1983) and Heikes (1983).

5.2. DISTRIBUTIONS AND CHEMICAL TRANSFORMATIONS OF NITROGEN COMPOUNDS

5.2.1. Ammonia

In the atmosphere, the chemistry of ammonia, a reduced form of nitrogen, appears to be largely independent of the oxidized nitrogen cycles. Ammonia is the only measured gaseous compound that acts as a base in clouds. (Its amine homologues are the only other known basic atmospheric gases.) Consequently, ammonia is important in determining the pH of fogs, clouds, and rain (Charlson and Rodhe 1982 and section 5.3.2.2). The main sinks for NH_3 appear to be removal by rainfall and into acid-aerosol particles.

The low oceanic concentrations of ammonia in marine air reported by Ayers and Gras (1983) of 15 to 100 pptv are consistent with the ratios of ammonium to excess sulfate ions found by Charlson in aerosols on the

Washington coast of 0.6 to 1.0 with median values of about 0.8 (unpublished data available from Dr. R. J. Charlson, University of Washington,
Seattle). The data of Duce (1983 and unpublished data available from Dr.
R. A. Duce, University of Rhode Island, Kingston) generally agree. Galloway et al. (1982) have found similar central values, ranging from 0.3
to 2.0, for ion ratios in marine rain, which presumably scavenges this
aerosol as well as ammonia and freshly formed sulfate (see Chapter 8).
Ratios corresponding to neutralized sulfuric acid or excess NH_4^+ are
found in less than 5% of their samples. Because ammonium is somewhat
better correlated with nitrate in these data sets than with sulfate,
these compounds may share some common source regions, possibly
continental.

Remote continental concentrations of ammonia are less well established than remote marine concentrations. The values from the data
available at the time of the Bermuda workshop probably reflected the high
variability of biological activities. Logan (1983) references some observations, mostly made near the detection limit of the instrument being
used, that relate to the typical continental boundary layer.

Ammonia is not expected to contribute much to the sources or sinks
of nitrogen oxides in the atmosphere and oxidation by OH radicals is
slow. The reaction

$$HO + NH_3 \longrightarrow NH_2 + H_2O \tag{5-7}$$

has a lifetime of 35-70 days if a day's 24-hour-averaged OH concentration
of $1-2 \cdot 10^6$ molec/cm^3 is assumed. NO_2 is also consumed by OH radicals.
Even if this reaction of NH_3 led to NO_2 exclusively, the most NO_2 lost by
reaction with OH radicals that could be replaced from NH_3 would be 1.5%.
This conversion does not appear to be important for the cycles of either
ammonia or the nitrogen oxides. However, amines react much more rapidly
with OH (Atkinson et al. 1977) and their fate may be the production of
other N compounds.

The reaction of NH_3 with HNO_3, also, does not appear to be important, except in polluted areas that have high concentrations of both
reactants. However, because of the reversibility of the reaction, we
agreed that it is probably not a major sink for either species.

5.2.1.1. Open Questions. We did not know the vertical or horizontal
distributions of NH_3 in remote or polluted areas at the time of the
workshop. More measurements of NH_3 concentrations are urgently needed,
particularly over the remote continental areas. Removal processes must
be better defined.

5.2.2. The Nitrogen Oxides

Most oxides of nitrogen can be interconverted into one another by chemical or photochemical reactions. (The pervasive but tropospherically
unreactive compound N_2O, a significant exception, was not further discussed at the workshop.) These compounds are commonly subdivided into
groups according to the time scales involved in their interconversions.
To clarify these time scales of interconversion and loss, we proposed the
following group classifications (Fig. 5-1, Table 5-1):

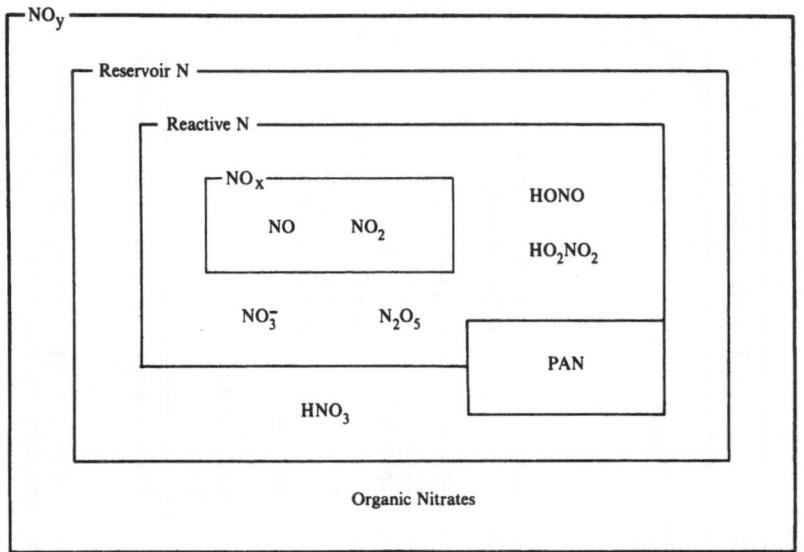

Figure 5-1. Classification of N Compounds by Reaction Rate. Reaction time scales increase from the center to the perimeter.

5.2.2.1. **Group I:** $NO_x = NO + NO_2$. Interconversion in this group is the fastest, occurring by the processes

$$NO + O_3 \longrightarrow NO_2 + O_2 \quad \text{and} \tag{5-8}$$

$$NO_2 + h\nu \longrightarrow NO + O, \tag{5-9}$$

followed by the rapid recombination of the O atoms to reform O_3. For 40 ppbv of O_3 in the clean troposphere, the conversion time is on the order of 1 to 2 minutes in daytime.

5.2.2.2. **Group II:** Reactive N ($NO + NO_2 + HONO + HO_2NO_2 + NO_3 + N_2O_5$). The interconversions between these compounds are on time scales of several hours. The two acids are formed by reactions with HO_x,

$$NO + OH \longrightarrow HONO \quad \text{and} \tag{5-10}$$

$$NO_2 + HO_2 \longrightarrow HO_2NO_2 \text{ (pernitric acid)}, \tag{5-11}$$

and are reconverted to the NO_x compounds by:

$$HONO + h\nu \longrightarrow HO + NO \quad \text{and} \tag{5-12}$$

$$HO_2NO_2 \longrightarrow HO_2 + NO_2. \tag{5-13}$$

For $OH = 10^6$ molec/cm^3, the formation time of HONO is ~ 2 hr. Because the photolysis time of HONO is short, the HONO concentration in the daytime

Table 5-1. Groupings and Lifetimes of NO_x Interconversions and Transformations.

Group	Species	Approximations for		
		Interconversion	Transformation	Physical Loss
I	$NO_x = NO + NO_2$	1-2 minutes	I $\longleftrightarrow$ II: 2-10 hours	
II	HONO, HO_2NO_2, NO_3, N_2O_5	Temperature dependent, several hours		
III	HNO_3		I $\longrightarrow$ III: 1.5 days III $\longrightarrow$ 15 days	1-15 days (wet and dry deposition
IV	PAN		PAN formation effective only at upper & middle troposphere I $\longrightarrow$ IV: 20-30 days IV $\longrightarrow$ I: dependent on mixing to lower troposphere from middle and upper tropospheres	30 days Mixing to lower troposphere and thermal decomposition

Notes: NO_3: Chemical losses are possible since NO_3 might react to form species outside the numbers of the groups. PAN formation requires hydrocarbons.

Sources: For physical loss of HNO_3, see Levine and Schwartz 1982. PAN mixing from upper to lower tropospheres is calculated using eddy diffusion coefficients from Liu et al. (1984).

is less than 1% of that of NO_x. Similarly, the fast thermal dissociation of pernitric acid assures that its concentration is also small relative to NO_x.

NO_3 is formed by

$$NO_2 + O_3 \longrightarrow NO_3 + O_2 \qquad (5\text{-}14)$$

and in daytime is photodissociated mainly by

$$NO_3 + h\nu \longrightarrow NO_2 + O \qquad (5\text{-}15)$$

at a sufficiently fast rate to make NO_3 concentrations negligible in daytime. At night, NO_3 reacts further:

$$NO_3 + NO_2 \longleftrightarrow N_2O_5. \qquad (5\text{-}16)$$

The photodissociation of N_2O_5 is also sufficiently fast to prevent appreciable concentrations in daytime. NO_3 appears to have additional reactive sinks at night (see section 5.2.2.6.) so that substantial nighttime N_2O_5 production is likely to be restricted to locations with many ppb of reactive nitrogen.

5.2.2.3. Group III: Reservoir N = Reactive N + PAN + HNO_3. NO_2 is converted to nitric acid by reaction with OH,

$$OH + NO_2 + M \longrightarrow HNO_3 + M, \qquad (5\text{-}17)$$

with a time constant of about 1.5 days for an OH concentration of 10^6. It can be reconverted by photolysis,

$$HNO_3 + h\nu \longrightarrow NO_2 + OH, \qquad (5\text{-}18)$$

with a partial time scale of about 15 days. This long lifetime makes HNO_3 a possible source of NO_x in remote regions if the nitric acid is protected against rainout. This would only be effective above regions of active low-cloud levels.

5.2.2.4. Group IV: PAN (peroxyacetyl nitrate). PAN is formed from NO_2 and the oxidation products of light hydrocarbons. The best measured species are the alkanes C_2H_6 and C_3H_8. For typical Northern Hemisphere concentrations of 2 ppb of C_2H_6 and 0.2 ppb of C_3H_8, production time scales for these hydrocarbons are on the order of one month. Other, poorly measured, hydrocarbons may increase the PAN-production rate.

The lifetime of PAN against thermal dissociation depends strongly on temperature and, therefore, on altitude in the troposphere (Table 5-2). The slow reaction with OH (Wallington et al. 1984) is important in the upper troposphere. Typical total lifetimes are shown in Table 5-1.

Under conditions of (NO) $\geq$ (HO_2), the lifetime of NO_2 conversion to PAN can be shown to be on the order of one month for concentrations of 1 ppb of C_2H_6 at 4-km altitude (Ehhalt and Rudolph 1984, Rudolph and Ehhalt 1981) in the remote Northern Hemisphere (Singh and Haust 1981,

Table 5-2. Typical Lifetimes of PAN During Summer (45° N).

Altitude (z) (km)	Temperature* (° K)	Lifetime (days)
0	296	0.35
2	286	2.8
4	273	36
6	261	60
8	248	60

Crutzen 1979). PAN formation is effective at altitudes above 4 km if a
source of NO_x exists in the upper troposphere. Lightning production of
NO_x has been proposed to have the necessary source strength (unpublished
data available from Dr. D. Kley, Institut für Chemie, Jülich, FRG). The
transfer time scale of PAN from the upper troposphere to the lower should
be on the order of 30 days. This estimate is based on model calculations
where PAN is transported down to the lower troposphere and dissociated
there to generate NO_x. Other sources of the PAN precursors are found in
industrial and biomass burning. These sources suggest similar long-
distance transport of PAN.

5.2.2.5. Hypotheses.

1. The long-range transport of nitrogen is by HNO_3 and PAN.

HNO_3 and PAN have the longest lifetimes of the nitrogen species we considered and are, therefore, candidates for the long-range transport of
nitrogen to oceanic and other remote regions (McFarland et al. 1979,
Crutzen 1979). Particulate nitrate, which is known to reside in the 1-μm
to 6-μm range, will fall out by gravitational settling or will be removed
in rain and cannot transport nitrogen for very great distances. Therefore, we concluded that particulate nitrate in remote oceanic regions is
probably formed in situ from the HNO_3 precursor reacting with the basic
sea-salt aerosol. As the next section indicates, there may well be other
nitrates that greatly increase this transport.

2. Upper tropospheric tropical lightning is a source of NO_x.

Kley (unpublished data available from Dr. D. Kley, Institut für Chemie,
Jülich, FRG) proposes that upper tropospheric tropical lightning is a
major source of the background NO_2 observed in remote oceanic areas. His
argument is based on the yield of NO_x in typical cloud-to-ground flashes
of $4 \cdot 10^{26}$ molec/flash and satellite observations of lightning-flash
frequency. He shows that the amount of nitrogen annually produced above
10 km in the tropical zone is sufficient to account for the observed
deposition to remote ocean regions. We estimated that the lightning
source above 10 km is providing 1.5 Tg (N)/yr.

5.2.2.6. Open Questions. Perhaps the simplest tests of the adequacy of our knowledge of chemical mechanisms at the time of the Bermuda workshop involved those for reactions that are so fast that transport has little effect. Daytime NO and NO_2 interconvert with time scales of 100 seconds. Several attempts have been made to rationalize the measured NO_2 and NO concentrations to the other relevant chemical and physical data. The relation in terms of a photostationary-state ratio is

$$\frac{[NO_2]}{[NO]} = \frac{k_{O_3}[O_3] + k_{ROO} \cdot [ROO \cdot]}{J_{NO_2}}, \qquad (5-19)$$

which considers that NO is converted NO_2, predominantly by reaction with ozone and with peroxy radicals, ROO (ROO· is HOO· or organic peroxy radicals). We wrote a normalized NO_x ratio, which ought to be close to one according to current knowledge, as

$$P = \frac{[NO_2]}{[NO]} \left[\frac{k_{O_3}[O_3] + k_{ROO} \cdot [ROO \cdot]}{J_{NO_2}} \right]^{-1} \qquad (5-20)$$

However, this number is frequently measured to be 2 or 3 or more. Ritter et al. (1979), Parrish et al. (1984), and Fahey et al. (1983, 1984) have shown that the deviation can be quite large and depends on solar flux. The concentration of ROO· has been crudely estimated in experiments (Cantrell et al. 1984) but calculations suggest that it is rarely on the order 10^9 molec/cm^3, a concentration required to produce P values near 2. Such large values for ROO· also imply very large ozone-production rates. However, ozone, nitrogen oxides, and J values are relatively uniform and sampled with adequate frequency when large P values are observed. These P values suggest that there are unknown measurement problems or unknown reactions. These unknown reactions might be greatly affecting our understanding of reactive-N chemistry and ozone-production chemistry.

Another open question involved a possible missing reactive-nitrogen compound. Measurements at the NOAA Aeronomy Lab's Niwot Ridge site in the Colorado mountains show that there is a very significant discrepancy between the sum of all the identifiable reservoir nitrogen compounds and the measurement of total N in an air sample that can be reduced to NO—a quantity frequently called NO_y (Parrish et al. 1984, Fahey et al. 1984). This measurement does not include N_2, N_2O, amines, or NH_3. The discrepancy, which is frequently as large as reservoir nitrogen in clean continental areas, may be attributable to an unidentified compound. Organic nitrates could be responsible. It is clear that we may be missing an important nitrogen-bearing compound present in a clean atmosphere.

We also considered that the nighttime destruction of NO_3 is poorly understood. The reactions of the NO_3 radical with the various organics that have been measured proceed relatively rapidly (Winer et al. 1984). However, nighttime NO_3 concentrations do not appear to reach their predicted concentrations in a clean troposphere (Noxon 1983 and unpublished

data available from Dr. D. Perner, Max-Planck Institut, Mainz, FRG, and
Dr. U. Platt, Institut für Chemie, Jülich, FRG). Noxon (1983) ascribes a
lifetime of approximately 20 minutes to nighttime NO_3 (equivalent to a
steady-state concentration of NO_3 at night), which is approximately one
1/100 that of NO_2 and approximately 50 times shorter than that given by
the conversion we discussed earlier (section 5.2.2.2). An additional
reaction must occur that may constitute a sink for reactive N, possibly a
reaction of DMS + NO_3 in marine environments. Alternatively, the thermal
decomposition of NO_3 may produce these low levels (Johnston et al. 1985).

Another question that remained unanswered was, Are there significant
liquid-phase reactions that remove the reactive nitrogen in cloud or
aerosol solutions? In the reviews of gas-liquid interactions by Heikes
(1983) and Chameides (1984), no novel liquid-phase reactions of importance are apparent in a remote atmosphere with very low NO_x concentrations. Although HO_2NO_2, NO_3, N_2O_5, and HONO are known to react on contact with water surfaces, the concentrations appear to be too low to be
significant. However, we agreed that radical and ion chemistry in atmospheric solution is a rapidly developing field (Chameides 1984).

5.3. CHEMICAL TRANSFORMATIONS OF SULFUR COMPOUNDS

5.3.1. Reduced Sulfur Compounds

A variety of atmospheric products corroborate the laboratory data on the
reaction of $(CH_3)_2S$ with OH. Yields of SO_2 under experimental conditions
vary from 20% to 95% of the $(CH_3)_2S$ consumed (Niki et al. 1983, Hatekeyama et al. 1982, Grosjean 1984). However, these results are only general guidelines because practicable laboratory experiments (1) require
much larger reactive nitrogen (~100 ppbv) concentrations than are found
in a remote troposphere (~ 50 pptv) and (2) do not usually provide us
with the means for understanding how the reaction of all intermediate
products may eventually produce SO_2.

Many compounds observed in the laboratory do not play a significant
role in tropospheric $(CH_3)_2S$ chemistry. Compounds like CH_3SNO, CH_3SONO,
and CH_3SONO_2 have been identified by or implicated from laboratory measurements (Niki et al. 1983, Grosjean 1984). These compounds appear to
be formed in the presence of high NO_x concentrations in the reaction
chambers and may not be typical of the remote atmosphere. High NO_x is
typically required to maintain large radical concentrations, to counter-
act wall effects, and to oxidize significant quantities of sulfide.
Grosjean (1984) reports conditions under which 90%-95% of reacted $(CH_3)_2S$
does form SO_2--arguably, these conditions are closest to atmospheric
ones.

The reaction of NO_3 with $(CH_3)_2S$ may contribute significantly. As
Atkinson et al. (1984) point out, the reaction of NO_3 with $(CH_3)_2S$ is so
fast ($k = 5.4 \cdot 10^{-13}$ cm^3/[molec $\cdot$ sec]) that NO_3 may provide an important contributory sink for the sulfide. The results of Tuazon et al.
(1984) suggest that the experiments of Atkinson et al. (1984) may allow
reaction rates as fast as 10^{-12} cm^3/(molec $\cdot$ sec). Nighttime concentrations of NO_3 as low as 1 pptv to 2 pptv may be as significant as OH at
average daytime concentrations of $2 \cdot 10^6$ molec/cm3. In simulations

using known reaction rates, these concentrations are easily attainable
from background NO_x (Logan 1983). However, NO_x concentrations of
100 pptv to 200 pptv would be required if, as Noxon (1983) suggests, NO_3
establishes a steady-state concentration of $[NO_3] \cong 0.01 [NO_2]$.

Some nighttime decay of dimethylsulfide is required. Measurements
of $(CH_3)_2S$ may vary diurnally with a ratio of maximum to minimum of ~ 1.6
(Andreae et al. 1985, see also Chapter 1). These variations are consistent with daytime concentrations of NO_x ranging from 10 pptv to 200 pptv,
depending primarily on the steady-state ratio of NO_3 to NO_2 at night.

Concerning H_2S and CS_2 reactions, Cox and Sheppard (1980) and Jones
et al. (1984) suggest that HO reactions give SO_2 with nearly unit yield
per molecule consumed. Jones and his colleagues roughly estimate the
yield of CS_2 photolysis in the presence of O_2. Chatfield and Crutzen
(1984) and Chatfield (1982) suggest that these reactions may lead to only
small incremental concentrations in a remote atmosphere. CS_2 may produce
from 8 pptv SO_2 in the lower troposphere to 15 pptv SO_2 in the upper
troposphere. According to our knowledge at the time of the Bermuda
workshop, more SO_2 production would be associated with an unacceptably
large production of COS.

5.3.1.1. Open Questions. H_2S and other sulfides may be transported from
continental sources (rain forests, wetlands) (Chatfield 1982) or even
volcanoes (Herrmann and Jaeschke 1984) to produce some oceanic SO_2.
Chatfield (1982) suggests that up to 15 pptv SO_2 may be present in the
boundary layer and 60 pptv SO_2 at 6-km altitude after substantial oceanic
trajectories. These simulations are highly uncertain, and concentrations
of 1/4 to 1/2 the quoted values are easily possible. Atmospheric H_2S
should be measured from airplanes high above continental areas (see
Delmas et al. 1980).

Compounds, such as dimethyl sulfoxide, dimethyl sulfone, methanesulfonic acid, and dimethylsulfide hydroperoxide,

```
            S
       H / \ H
      HC      C - OOH ,
       H      H
```

should be measured more intensively (see Saltzman et al. 1983 for
methanesulfonic acid and Andreae 1980 for dimethyl sulfoxide). These
compounds are indicated as likely oxidation products (Niki et al. 1983,
Grosjean 1984, Atkinson et al. 1984) but are apparently seen at very
small concentrations compared to SO_2. Even if they are only minor oxidation products, they might be very useful tracers for the origin of SO_2.

5.3.2. Sulfur Dioxide

Although it is well established that SO_2 reacts in the atmosphere to
produce both sulfate aerosol particles and sulfate ions in cloud water,
we do not know, with much accuracy, the rates of reaction, the factors
controlling the rates, or the relative importance of the various parallel
reaction pathways. We believed the following hypotheses should be
tested.

5.3.2.1. Hypotheses.

1.   SO₂ oxidation does proceed significantly by OH radical oxida-
     tion in clean air.

Model calculations by Rodhe and Isaksen (1980) and Chatfield and Crutzen
(1984) show that up to 25 mg(S)/(m² · yr) of SO₂ removed from the atmosphere can proceed by the reaction OH + SO₂ + M. These calculations
depend only on the measured remote, oceanic distribution of SO₂ (Maroulis
et al. 1980), the relatively well-measured rate coefficient (Baulch et
al. 1982), and the globally averaged OH-radical concentrations necessary
to explain the tropospheric balances of many carbon and chlorine species
(Logan et al. 1981). This represents around 40% of the removal of SO₂ in
the simulations; dry deposition and liquid-phase reaction are required
for roughly commensurate shares of the total removal process.

2.   Liquid-phase oxidation apparently proceeds significantly in a
     clean troposphere.

Several mechanisms are known to operate rapidly in the laboratory, some
of which are discussed below.

5.3.2.2.   Open Questions: Liquid-phase Oxidation. Our main questions
regarding SO₂ oxidation in remote tropospheric clouds probably concerned
the transport time of an SO₂ molecule to a cloud. This meteorological
transport may be a limiting rate (see Part III).
     Other open questions revolved around physical inhibitions. Schwartz
(1983) has recently reviewed several restrictions to transport that may
hinder absorption and the reaction of extremely reactive gases. Resistances in the air, at the (cloud or aerosol) droplet surface and within
the droplet, must be considered. Laboratory reaction rates that are fast
(time scales of minutes) must be checked. The equations of transport and
reaction must be solved simultaneously outside and within the droplet.
Even species like HO₂ may establish a Henry's Law equilibrium near the
surface of the droplet because of restricted transport to the interior.
     Chemical inhibitions also presented some unanswered questions. The
formaldehyde-SO₂ adduct forms at a rapid rate in solution (Jacob and
Hoffman 1983). Measured formaldehyde concentrations in a remote marine
mixed layer are 200-300 ppt (Lowe et al. 1980). The availability of
other hydrocarbon sources suggests that it may be higher in vegetated
continental regions. Some of the oxidation reactions, notably the ozone
reaction, are inhibited by [H⁺] and, therefore, by H₂SO₄ production
(Taylor et al. 1983). The remote atmosphere exhibits a wide variety of
pH values for rain and cloud water due to three factors. Two are
straight forward, the third is less appreciated:

1.   The liquid-water content of clouds and the intensity of
     rain vary widely, thereby allowing for a range of dilution of
     dissolved anions by more than a factor of 10 (Clarke and Ber-
     barian 1984).

2.   Inputs to dissolved anions also vary depending on their precur-
      sors in gas and aerosol phases. These, in turn, depend on
      sources and sinks that vary geographically and temporally
      (Charlson and Rodhe 1982). Such variations can strongly in-
      fluence some significant reactions, such as the oxidation of S
      (IV) by ozone.

3.   The typical molar ratios of ammonium and sulfate in cloud water
      and rainwater are often in a region of $(NH_4^+) \sim (SO_4^=)$ where
      ammonia and $SO_4^=$ both have strong effects on pH (see Chapters 1
      and 8).

Figure 5-2 shows calculations of pH values expected in a cloud of
0.5 g/m^3 liquid water as a function of both total ammonia and sulfate
aerosol concentration (Vong and Charlson 1985). It is clear that pH is
extremely sensitive to both NH_3(total) and $SO_4^=$ aerosol under background
conditions of ~0.1 μg/m^3 NH_3(total) and a fraction of a μg/m^3 $SO_4^=$.
Thus, if any reaction rate in cloud water is pH dependent, we must expect
the levels of both NH_4^+ and $SO_4^=$ to have control.

The availability of oxidants was also not resolved. Hydrogen peroxide is thought to reach the 1 ppb level in a remote clean atmosphere,
perhaps higher in tropical regions (Logan et al. 1981, Crutzen and Gidel
1983, Chatfield and Crutzen 1984). Measurements in apparently clean
rainwater of around 1 ppm (Kok 1981, Zika et al. 1982) concur with
simulated results and establish the presence of enough H_2O_2 to oxidize S
(+4). More measurements of hydrogen peroxide are required, especially in
the tops of clouds where significant quantities of H_2O_2 can be formed by
radical and ionic reactions involving HO_2 (Chameides 1984). Ozone is
almost always available in great excess and the simplest calculations
suggest it can oxidize SO_2 very rapidly near and above pH ~ 5.5.

5.3.2.3. Open Question: Radical Oxidation in Clouds. There appears to
be sufficient oxidants for S (+4) in remote cloud water in clean areas,
provided inhibition does not occur. However, radical and redox reactions
may contribute to more direct oxidation or more peroxide production.
They may also alter the oxidation in as yet unperceived ways. (See
Chameides [1984] for a recent review particularly applicable to remote
marine clouds.)

5.4. PHYSICAL PROCESSES IN AEROSOL PARTICLES AND CLOUDS

Several physical processes are listed in Figure 4-4, some of which are
closely involved in controlling the overall lifetime of key S and N
species. Some time scales for physical processes are comparable in
magnitude to those for chemical processes. The three processes by which
S-containing gases (including SO_2) can be transformed into accumulation-
mode $SO_4^=$ are by (1) the formation of new $SO_4^=$ particles with coagulation
to larger sizes, (2) the deposition on preexisting aerosol particles of
gas-phase reaction products, (3) the dissolution into cloud water, reaction in cloud, and evaporation back to accumulation-mode aerosol.

We hypothesized that process 1 must occur if aerosol-number concentrations are to be maintained, but there was a consensus that it does not

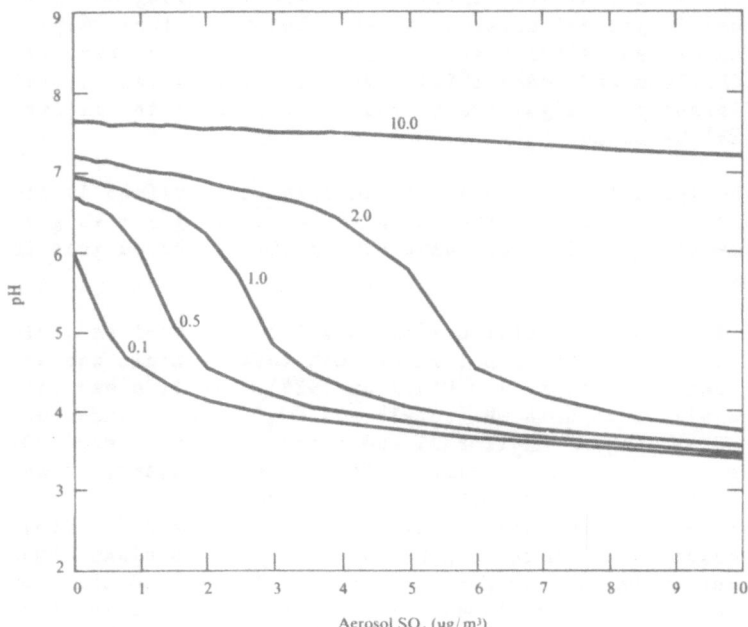

Figure 5-2. pH Sensitivity to $SO_4^=$ and NH_4^+: Model calculations of expected pH of cloud water or rainwater for cloud-liquid-water content
of 0.5 g/m^3, 100 pptv of SO_2, 330 ppm CO_2, and no NO_3^-. The ab-
scissa shows the assumed input of aerosol sulfate in μg/m^3 and the
ordinate shows the calculated equilibrium pH. Each line corresponds
to the indicated amount of total NH_3 + NH_4^+ in units of μg/m^3 of
cloudy air (Vong and Charlson 1985).

represent the main pathway of $SO_4^=$ production. The choice between processes 2 and 3 was open. The fluxes that are likely to be of similar
magnitude must depend on climate (e.g., cloudiness).

5.4.1. Aerosol Process Lifetimes

What are the aerosol lifetimes in meteorologically or climatologically
different locations, particularly in the upper troposphere? Aerosol
lifetimes have been estimated in only a few publications (see Table 5-3).
These times are clearly comparable to some of those for chemical
processes.

We hypothesized that the overall lifetime of NH_3 species, oxidized
sulfur species, and HNO_3 species can be governed in some situations by
aerosol physical processes. As long as total ammonia is not present in
excess of acidic sulfate, its physical pathways are determined mainly by
acidic sulfate in aerosol or in clouds. In contrast, HNO_3 appears either

Table 5-3. Aerosol Particle Lifetimes

Process	Lifetime	Reference
Coagulation of nuclei-mode w/accumulation-mode particles	10–200 hr: depending on number population of accumulation mode	Ogren & Charlson 1983
Removal of soluble aerosol by boundary layer processes	1 day: mid-lat., winter 4 days: mid-lat, summer	Ogren & Charlson 1983 Rodhe & Grandell 1981
Global tropospheric aerosol removal, all processes (based on radioactivity studies)	10–14 days: global	Junge 1963, Jaenicke 1980
Removal of coarse particles near surface by sedementation	Few hours	

Note: Compare to Table 7-2. Estimates of lifetimes for transport of
       sulfur gases and aerosol into clouds and into raining clouds are
       given in the Part III (Chapters 6 and 7).

to be dissolved in cloud droplets or raindrops or to be adsorbed on the
surface of basic salt or soil dust. Different lifetimes for $SO_4^=$ and
NO_3^- are indicated.

5.4.2. Open Questions

Because these physical processes are consecutive with the initial chemical reactions of key S and N species and because they are competitive
with dry removal processes to the surface, the question remains as to the
overall lifetime under prescribed atmospheric conditions.

5.5. CONCLUSIONS AND RECOMMENDATIONS

- Major pathways for the physical and chemical transformations
  appear to be understood qualitatively and agreed upon. However, it is certainly possible that important new pathways may
  be discovered.

- Rates of processes should be quantified.

- Simultaneous observations should be taken to measure all or
  most relevant species and to determine their physical states.
  Tests must be devised to allow the observers to follow the

atmospheric life of an atom of sulfur or nitrogen from its
source to its sink. This would necessarily require an understanding of both the chemical and the physical transformation
processes (including meteorological processes).

- Laboratory studies should be conducted to examine individual
  processes under a realistic range of chemical and physical
  conditions. Field observations should be compared to test
  hypotheses.

- A variety of compounds, e.g., dimethyl sulfoxide and organic
  nitrates, that are suspected should be identified and
  quantified.

- The existence of unsuspected compounds should be explored by
  assaying the total sulfur and non-N_2 in nitrogen in air samples
  and all the compounds that contribute to the totals should be
  identified.

5.6. REFERENCES

Andreae, M. O. 1980. Dimethylsulfoxide in marine and freshwaters.
    Limnol. Oceanogr. 25:1054-1063.
Andreae, M. O., R. J. Ferek, F. Bermond, K. P. Byrd, R. T. Engstrom, S.
    Hardin, P. D. Houmere, F. LeMarreck, R. B. Chatfield. 1985. Di-
    methylsulfide in the marine atmosphere. J. Geophys. Res. (in
    press).
Atkinson, R., and A. C. Lloyd. 1984. Evaluation of kinetic and mecha-
    nistic data for modeling of photochemical smog. J. Phys. Chem. Ref.
    Data 13:315-444.
Atkinson, R., R. A. Perry, and J. N. Pitts. 1977. Rate constants for
    the reaction of the OH radical with CH_3SH and CH_3NH_2 over the tem-
    perature range 299-426 K. J. Chem. Phys. 66:1578-1580.
Atkinson, R., J. N. Pitts, Jr., and S. M. Aschmann. 1984. Tropospheric
    reactions of dimethyl sulfide with NO_3 and OH radicals. J. Phys.
    Chem. 88:1584-1587.
Ayers, G. P., and J. L. Gras. 1983. The concentration of ammonia in
    southern ocean air. J. Geophys. Res. 88:10,665-10659.
Baulch, D. L., R. A. Cox, P. J. Crutzen, R. F. Hampson, Jr., J. A. Kerr,
    J. Troe, and R. T. Watson. 1982. Evaluated kinetic and photochemical
    data for atmospheric chemistry: Supplement I. J. Phys. Chem. Ref.
    Dat. 11:327-496.
Baulch, D. L., R. A. Cox, R. F. Hampson, Jr., J. A. Kerr, J. Troe, and R.
    T. Watson. 1984. Evaluated kinetic and photochemical data for
    atmospheric chemistry: Supplement II. J. Phys. Chem. Ref. Dat. 13:
    1259-1380.
Cantrell, C. A., D. H. Stedman, and G. J. Wendell. 1984. Measurement of
    atmospheric peroxy radicals by chemical amplification. Anal. Chem.
    56:1496-1502.
Chameides, W. 1984. The photochemistry of a remote marine stratiform
    cloud. J. Geophys. Res. 89:4739-4756.

Charlson, R. J., and H. Rodhe. 1982. Factors controlling the acidity of rainwater. Nature 295:683-685.

Chatfield, R. B. 1982. Remote tropospheric SO_2: Cloud transport of reactive sulfur emissions. NCAR Rept. No. CT-70, Natl. Center for Atmos. Res., Boulder, CO.

Chatfield, R. B., and P. J. Crutzen. 1984. Sulfur dioxide in remote tropospheric air: Cloud transport of reactive precursors. J. Geophys. Res. 89:7111-7132.

Clarke, A. D., and G. Berbarian. 1984. Rapid variability in precipitation chemistry for individual tropical marine cumulus. Presented at Am. Geophys. Union Fall Meeting, San Francisco, CA. Dec. (Paper available from Dr. Clarke, University of Washington, Seattle.)

Cox, R. A., and D. Sheppard. 1980. Reactions of OH radicals with gaseous sulphur compounds. Nature 284:330-331.

Crutzen, Paul J. 1979. The role of NO and NO_2 in the chemistry of the troposphere and stratosphere. Ann. Rev. Earth Planet. Sci 7:443-472.

Crutzen, P. J., and L. T. Gidel. 1983. A two-dimensional photochemical model of the atmosphere: 2. The tropospheric budgets of the anthropogenic chlorocarbons, CO, CH_4, CH_3Cl, and the effects of various NO_x sources on tropospheric ozone. J. Geophys. Res. 88:6641-6661.

Delmas, R., J. Baudet, J. Servant, and Y. Baziard. 1980. Emissions and concentrations of hydrogen sulfide in the air of the tropical forest of the Ivory Coast and of temperate regions in France. J. Geophys. Res. 85:4468-4474.

Duce, R. A. 1983. Biogeochemical cycles and the air-sea exchange of aerosols. In SCOPE 24: The Major Biogeochemical Cycles and Their Interactions (B. Bolin and R. B. Cook, eds.), New York:Wiley.

Ehhalt, D. H., and J. Rudolph. 1984. On the importance of light hydrocarbons in multiphase atmospheric systems. Ber. der Kernforschung. Jül. 1942, Jülich, West Germany.

Fahey, D. W., D. D. Parrish, J. M. Roberts, S. C. Liu, D. L. Albritton, F. C. Fehsenfeld. 1983. Photochemical oxidants in the rural troposphere. Presented at the Am. Geophy. Union Fall Meeting, San Francisco, CA., Dec. (Paper available from Dr. Fahey, NOAA, Boulder, CO.)

Fahey, D. W., G. Huebler, E. J. Williams, P. C. Murphy, R. B. Norton, D. L. Albritton, and F. C. Fehsenfeld. 1984. Correlation of NO_y with NO_x, HNO_3, and O_3 in the rural troposphere. Presented at the Am. Geophys. Union Fall Meeting, San Francisco, CA., Dec. (Paper available from Dr. Fahey, NOAA, Boulder, CO.)

Galloway, J. N., G. E. Likens, W. C. Keene, and J. M. Miller. 1982. The composition of precipitation in remote areas of the world. J. Geophys. Res. 87:8771-8786.

Grosjean, D. 1984. Photooxidation of methyl sulfide, ethyl sulfide, and methanethiol. Environ. Sci. Technol. 18:460-468.

Hatekeyama, S., M. Okuda, and H. Akimoto. 1982. Formation of sulfur dioxide and methane sulfonic acid in the photooxidation of dimethyl sulfide in air. Geophys. Res. Let. 9:583-586.

Heikes, B. 1983. The chemistry of acid generation in the troposphere:
    The solution phase and heterogeneous processes. In Regional Acid
    Deposition: Models and Physical Process (J. G. Wyngaard, Project
    Director). NCAR Proj. Rept. No. NCAR/TN-214+STR, Natnl. Center
    Atmos. Res., Boulder, CO., 154-224.
Herrmann, J., and W. Jaeschke. 1984. Measurement of H_2S and SO_2 over
    the Atlantic Ocean. J. Atmos. Chem. 1:111-123.
Jacob, D. J., and M. R. Hoffman. 1983. A dynamic model for the produc-
    tion of H^+, NO_3^-, and $SO_4^=$ in urban fog. J. Geophys. Res. 88:
    6611-6621.
Jaenicke, R. 1980. Atmospheric aerosols and global climate. J. Aerosol
    Sci. 11:577-588.
Johnston, H. S., C. A. Cantrell, and J. G. Calvert. 1985. Unimolecular
    decomposition of NO_3 to form NO and O_3. J. Phys. Chem. (in press).
Jones, B. M. R, R. A. Cox, and S. A. Penkett. 1984. Atmospheric chemis-
    try of carbon disulphide. Atmos. Chem. 1:53-64.
Junge, C. E. 1963. Air Chemistry and Reactivity. New York:Academic Press.
Kok, G. L. 1981. Measurements of hydrogen peroxide in rainwater. Atmos.
    Environ. 14:653-656.
Levine, S. Z., and S. E. Schwartz. 1982. In-cloud and below-cloud
    scavenging of nitric acid vapor. Atmos. Environ. 16:1725-1734.
Liu, S. C., J. R. McAfee, and R. J. Cicerone. 1984. Radon 222 and
    tropospheric vertical transport. J. Geophys. Res. 89:7285-7295.
Logan, J. A. 1983. Nitrogen oxides in the troposphere, global and re-
    gional budgets. J. Geophys. Res. 88:10,785-10,807.
Logan, J. A., M. J. Prather, S. C. Wofsy, and M. B. McElroy. 1981.
    Tropospheric chemistry: A global perspective. J. Geophys. Res.
    86:7210-7254.
Lowe, P. C., U. Schmidt, and D. H. Ehhalt. 1980. A new technique for
    measuring formaldehyde. Geophys. Res. Let. 7:825-828.
Maroulis, P. J., A. L. Torres, A. B. Goldberg, and A. R. Bandy. 1980.
    Atmospheric SO_2 measurements on project GAMETAG. J. Geophys. Res.
    85:7345-7349.
McFarland, M., D. Kley, J. W. Drummond, A. L. Schmeltekopf, and R. J.
    Winkler. 1979. Nitric oxide measurements in the Equatorial Pacific.
    Geophys. Res. Let. 6:605-608.
Niki, H., P. D. Maker, C. M. Savage, and L. P. Breitenbach. 1983. An
    FTIR study of the mechanism of the reaction HO + CH_3SCH_3. Int. J.
    Chem. Kinet. 15:647-654.
Noxon, J. F. 1983. NO_3 and NO_2 in the mid-Pacific troposphere. J.
    Geophys. Res. 88:11,017-11,021.
Ogren, J. A., and R. J. Charlson. 1983. Elemental carbon in the atmo-
    sphere: cycle and lifetime. Tellus 35B:241-254.
Parrish, D. D., R. B. Norton, D. W. Fahey, G. Häbler, P. D. Goldan, D. L.
    Albritton, and F. C. Fehsenfeld. 1984. Correlations among tropo-
    spheric species measured at Niwot Ridge, Colorado. Presented at Am.
    Geophys. Union Fall Meeting, San Francisco, Dec. (Paper available
    from Dr. Parrish, NOAA, Boulder, CO 80303.)
Ritter, J. A., D. H. Stedman, and T. J. Kelly. 1979. Ground-level
    measurements of nitric oxide, nitrogen dioxide, and ozone in rural
    air. In Nitrogeneous Air Pollutants (D. Grosjean, ed.), Ann Arbor:
    Ann Arbor Science, 325-343.

Rodhe, H., and J. Grandell. 1981. Estimates of characteristic times for precipitation scavenging. J. Atmos. Sci. 38:370–386.

Rodhe, H., and I. Isaksen. 1980. Global distribution of sulfur compounds in the troposphere estimated in a height/latitude transport model. J. Geophys. Res. 85:7401–7409.

Rudolph, J., and D. H. Ehhalt. 1981. Measurements of C_2–C_5 hydrocarbons over the North Atlantic. J. Geophys. Res. 86:11,959–11,964.

Saltzman, E. S., D. L. Savoie, R. G. Zika, and J. M. Prospero. 1983. Methane sulfonic acid in the marine atmosphere. J. Geophys. Res. 88:10,897–10,902.

Schwartz, S. E. 1983. Mass–transport considerations pertinent to aqueous-phase reactions of gases in liquid–water clouds. Presented at the NATO Advanced Study Institute on Chemistry of Multiphase Systems. NTIS No. BNL–34174. (Available from Dr. Schwartz, Brookhaven Natnl. Lab., New York.)

Singh, H. B., and P. L. Haust. 1981. Peroxy acetyl nitrate (PAN) in the unpolluted atmosphere: An important reservoir for nitrogen oxides. Geophys. Res. Lett. 86:941–944.

Taylor, G. S., M. B. Baker, and R. J. Charlson. 1983. Heterogeneous interactions of the C, N and S cycles in the atmosphere: The role of aerosols and clouds. In SCOPE 24:The Major Biogeochemical Cycles and Their Interactions (B. Bolin and R. B. Cook, eds.), New York: Wiley, 115–124.

Tuazon, E. C., E. Sanhuseza, R. Atkinson, W. P. L. Carter, A. M. Winer, and J. N. Pitts, Jr. 1984. Direct determination of the equilibrium can start at 298K for the NO_2 + NO_3 $\rightleftharpoons$ N_2O_5. J. Phys. Chem. 68:3095–3098.

Wallington, T. J., R. Atkinson, and A. M. Winer. 1984. Rate constants for the gas phase reaction of OH radical with peroxy acetyl nitrate (PAN) at 273 and 297 K. Geophys Res. Lett. 11:861–864.

Weston, R. E., and H. A. Schwartz. 1972. Chemical Kinetics. Englewood Cliffs, New Jersey: Prentice Hall.

Winer, A. M., R. Atkinson, and J. N. Pitts. 1984. Gaseous nitrate radical: Possible atmospheric sink for biogenic organic compounds. Science 224:156–159.

Vong, R. J., and R. J. Charlson. 1985. The equilibrium pH of a cloud or raindrop: A computer-based solution for a six-component system. J. Chem. Ed. 62:141–143.

Zika, R., E. Saltzman, W. L. Chameides, and D. D. Davis. 1982. H_2O_2 levels in rainwater collected in South Florida and the Bahama Islands. J. Geophys. Res. 87:5015–5018.

PART III

THE TRANSPORT OF SULFUR AND NITROGEN THROUGH THE REMOTE ATMOSPHERE

6. THE TRANSPORT OF SULFUR AND NITROGEN THROUGH THE REMOTE ATMOSPHERE BACKGROUND PAPER

Henning Rodhe
Department of Meteorology
Arrhenius Laboratory
University of Stockholm
S-10691 Stockholm, Sweden

6.1. INTRODUCTION

The emphasis of this paper is on reviewing previous studies of the atmospheric transport processes affecting sulfur and nitrogen compounds. More future-oriented perspectives are in Chapter 7.

Studies of air pollutants have taught us that the concentration of gases and particulate matter may be very substantially elevated in air close to major sources (e.g., industries and urban areas). Such local influences may also occur around significant natural sources, such as tidal flats or fumaroles. Precipitation samples are normally less influenced by nearby sources since much of the material is incorporated in hydrometeors at cloud levels. However, solid dust and vegetation debris may cause significant local contamination particularly when bulk collectors are used (Galloway and Likens 1978). The possibility of a very local influence must be considered when interpreting measurements from remote areas.

This paper does not focus on short-range transport (tens of kilometers) but rather deals primarily with larger transport scales that range from hundreds to many thousands of kilometers—i.e., from long-range transport of a hundred to a few thousand kilometers to very-long-range transport of several thousand kilometers to global transport of many thousands of kilometers. Sometimes the term long-range transport is also used to encompass all these spatial scales.

Compared to transport on shorter scales (of tens of kilometers), the following differences become important in studies of long-range transport.

1. Air parcels do not necessarily move in straight lines along the prevailing mean wind. Because of variations of winds in space and time, air trajectories have a more complex structure.

2. A deeper layer of the atmosphere is involved. At time scales longer than a few days (corresponding to transport scales $> 10^3$ km), a significant part of the transport may take place in the free troposphere even if sources (or sinks) are within the planetary boundary layer (PBL).

J. N. Galloway et al. (eds.), The Biogeochemical Cycling of Sulfur and Nitrogen in the Remote Atmosphere, 105–124.
© 1985 by D. Reidel Publishing Company.

3.   Chemical-transformation and physical-removal processes often
      have a decisive influence on concentrations. As a consequence,
      transport, transformation, and removal must be considered
      simultaneously.

    The subsequent parts of this paper are devoted to a general discus-
sion about the processes of long-range to global-scale transport and
about the techniques used to identify source-receptor relationships. A
selection of case studies is also included.

6.2. METHODS OF STUDYING SOURCE-RECEPTOR RELATIONSHIPS

The receptor-oriented transport problem is the problem of identifying the
source or sources of an element or a compound that has been found in
remote areas. The standard techniques to do this include studies of air-
parcel movements (trajectories), information on other species measured
simultaneously (tracers), and mathematical-transport models. These techniques are also being used to study the dispersion of material from a
known source--the source-oriented transport problem.

6.2.1. Air-parcel Movements

A well-known technique being used to obtain a bearing to the source by
simultaneous measurements of the concentration in the air and the direction of the wind was established in studies of urban air-pollution problems. To identify not only the direction of the source but also the
distance to it requires at least two measurement sites that are placed
far enough apart to distinguish any difference in direction.
    For long-range transport covering hundreds of kilometers or more,
the local wind direction is often a poor measure of transport direction
and proper trajectories have to be considered. Rodhe et al. (1972) used
isobaric trajectories to identify the major sources of soot and sulfate
aerosols in southern Sweden. At that time, we found that the sources are
both on the European continent and in the United Kingdom, as far away as
thousands of kilometers. Similar studies have since been made for several different pollutants from the industrialized regions of the world,
usually where there was an a priori knowledge of source characteristics
(point or area, elevated or surface, industrial or urban). In remote
locations, the usefulness of air trajectories for tracing air-mass movement is often limited by a lack of reliable wind data.
    In those cases where explicit trajectories are not available, other
characteristics of the air mass--e.g., temperature, moisture, and visi-
bility--are sometimes used as indicators of the source regions of an
observed trace species. The problem of calculating air trajectories from
meteorological data has been discussed by several authors (cf. reviews in
Danielson 1961, Pack et al. 1978, Shannon and Patterson 1983, Eliassen
1984). A few of their more important points are summarized below.
    The precise definition of an air trajectory is somewhat arbitrary
since it depends on the choice of the scale separating the advective part
of the wind (the mean wind) from the smaller scale fluctuations (the
turbulence). In many applications, the separation scale is determined by
the grid distance of the model that is used to analyze the meteorological

data. For example, if the grid distance is on the order of 100 km, as is common in most large-scale dynamical models, no mesoscale motion (land-sea breeze, etc.) is resolved by the mean-wind variables.

Air movements are in reality three-dimensional. Since the vertical component of the wind is virtually impossible to determine from direct observations and very difficult to calculate indirectly, most calculated trajectories are forced to be horizontal (or isobaric). This simplification can introduce a very large error into the trajectory calculation if vertical movements are significant, as in the case of precipitation events. Therefore, great care must be exercised when using trajectories to identify the sources of contaminants found in precipitation samples.

Related to the previous point is the question of the exchange between the planetary boundary layer (PBL) and the free troposphere. In situations with a well-defined PBL with stable stratification or limited by a stable layer above (e.g., over continents in winter or when warm air is advected over a cooler ocean surface), the air is likely to remain within the PBL for several days. PBL trajectories can then be used to simulate transport over such time periods. Over land, when the daily cycle of surface heating causes the PBL to breathe, a substantial fraction of the air in the PBL is exchanged with tropospheric air (Sisterson et al. 1979). Material transported over long distances in the free troposphere may then contribute significantly to concentrations measured in the PBL. In such cases, 3-D trajectories originating at the top of the PBL over the measurement site may help to provide information on sources. In situations with deep convective motions (Cb clouds) or in connection with frontal systems, PBL trajectories are completely inadequate to trace air movements on time scales of days or longer.

Even aside from the problem of the vertical motion, all calculations of individual trajectories are bound to be uncertain because of inadequate meteorological data, poor interpolation procedures, etc. Trajectories over one or two days may be reasonably accurate in many situations. However, those covering more than 2 or 3 days are considered as no more than indications. In studies of individual situations, it is a good rule to base any judgment on several trajectories separated by small but finite distances in time and space. Only if the different trajectories agree reasonably well may any confidence be placed in the results. At many remote sites, in particular those in ocean areas, the meteorological data are insufficient to enable air trajectories to be calculated.

In climatological studies involving many trajectories, the uncertainty of individual trajectories is much less of a problem. Trajectory ensembles give reasonable estimates of the most common origin (on time scales of several days) of air masses affecting a particular site (Miller 1981a, b) or of dispersion from a particular source (Durst et al. 1959). However, systematic errors can still lead to biases in the trajectory statistics. A case in point is the difference between trajectories during precipitation and nonprecipitation periods. For example, Miller and Harris (1985) have found that the frequency of occurrence of trajectory directions at Bermuda differs substantially between rainy and dry days. It would clearly be misleading to use trajectory statistics based on all weather types to interpret the composition of rainfall samples.

For the midlatitudes of the Northern Hemisphere, except for parts of the oceans, the meteorological data available from the major meteoro-

logical centers are good enough to calculate large-scale, 3-D trajectories for a couple of days with reasonable accuracy. The situation is
much less satisfactory for the tropics and for the Southern Hemisphere.
In the latter parts of the world, the best data base available is probably the FGGE data for December 1978 through November 1979. Gridded data
on winds (including the vertical component), temperature, pressure
(mass), and moisture are available.
     Figure 6-1 illustrates some of the points mentioned above. It shows
four different trajectories arriving at Bermuda at 18Z on 14 November
1980. It is only for the first day that the agreement is reasonable.
The subsequent differences between the 2-D and the 3-D versions, both of
which originate at 850 mbar (B and C), are quite dramatic.

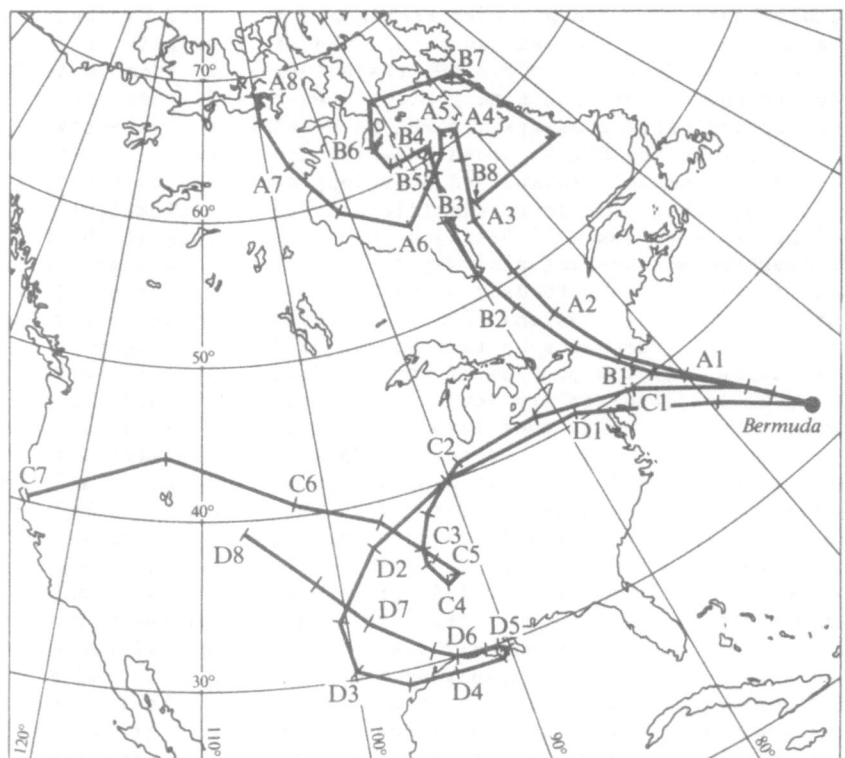

Figure 6-1. Receptor-oriented Trajectories Arriving at Bermuda on 14
    November 1980 at 18Z. A: 950-mbar isobaric, B: 850-mbar isobaric,
    C: 3-D arriving at 850 mbar, D: 3-D arriving at 700 mbar. Positions
    are marked with a dash for every 12 hours and with a figure indicat-
    ing the day. The height of the 850-mbar 3-D trajectory, C, varies
    from about 750 mbar on day 1 to 865 mbar on day 4 and the height of
    the 700-mbar 3-D trajectory, D, from 605 mbar on day 1 to 710 mbar
    on day 4.

6.2.2. Tracer Studies

Many different tracers have been used as indicators of the origin of
sulfur and nitrogen species themselves or of the air masses in which they
are measured. Some of them are mentioned below.

Pack and his colleagues (1977) used chlorofluorocarbons (CFC) as
tracers of industrial pollution at Adrigole, Ireland. CFC can at least
be used as a qualitative indication of whether the air has passed over an
industrial region during the preceding several days. Similar compounds,
including SF_6, can be used in field experiments where known amounts of
the tracer are emitted and the air movements are studied by downwind
measurements of the tracer. In most cases, such tracer experiments are
limited to spatial scales of less than a few tens of kilometers.

Other tracers of industrial pollution have been considered, including O_3, CO, and CO_2. The value of these is limited because their concentrations are being influenced by chemical transformations (O_3 and CO) and
by diffuse biological sources and sinks (CO, CO_2). Sometimes O_3 and CO
are used as indicators of the intrusion of stratospheric air (e.g.,
Seiler and Fishman 1981).

Elemental carbon (soot) is useful as a tracer of industrial activity
because it is chemically inert. Rahn et al. (1980) have used elemental
carbon to infer sources of aerosol particles in Arctic air. At lower
latitudes, forest and grass fires represent an additional source (Andreae
et al. 1984).

Because of the enrichment of potassium in plant tissues, the K content
of particulate matter is sometimes used as a tracer of plant material,
either from windblown debris or from fires (Andreae et al. 1984).

Na, Cl, or Mg are often used as a tracer of marine influence on
samples of aerosol particles and precipitation (Eriksson 1959, Junge
1963). Similarly, Al, Si, and Ca are being considered indicators, albeit
less specific, of the presence of soil dust (Rahn 1981).

Heavy metals and some other elements in aerosol particles are used
as tracers in studies of Arctic haze (see below). In particular, Rahn
and Lowenthal (1984) have used a combination of seven elements and suggest that signature ratios between these elements may be used as tracers
of specific industrial regions.

Radon is a radioactive gas emitted to the atmosphere from land
surfaces. Both the gas itself and its daughter products, in particular
^{210}Pb, can be used as tracers of air from the boundary layer over the
continents (Junge 1963).

Stable isotopes of sulfur, nitrogen, and oxygen are extensively used
by geochemists and other scientists to trace the origin or the track of
reactions of sulfur and nitrogen compounds. Compared to the previously
mentioned tracers, the isotopic signature has the advantage of tracing
the actual compounds of interest. (For reviews, see Nielsen 1974, Smith
1975, Krouse 1980, and for specific studies, see Ostlund 1959, Mizutani
and Rafter 1969, Nriagu and Coker 1978, Grey and Jensen 1972, Hitchcock
and Black 1974). Unfortunately, the isotopic composition of the different element reservoirs is not always very specific--for example, $\delta^{34}S$ of
coal varies from −3 to +27 (Thode and Rees 1970)--and large isotopic
fractionations take place in various chemical and physical transformation

processes. Many studies using this technique are, therefore, not quite
conclusive (cf., Krouse 1980). Nevertheless, the composition of the
stable isotopes represents a potentially important source of information
about sources of S and N compounds and the use of it—preferably in
conjunction with other tracers—should be encouraged.

6.2.3. Transport Models

To go beyond a simple trajectory analysis to ascertain the possible
sources of a measured contaminant, a quantitative model needs to be
formulated that will incorporate emission, transport, transformation, and
removal processes. Such a model might be based on trajectories (Lagrangian model) or on data specified at fixed grid points (Eulerian model).
Lagrangian long-range-transport (LRT) models are in principle simpler to
handle than the corresponding Eulerian models, i.e., cheaper to run on
the computer and less susceptible to numerical difficulties. On the
other hand, Eulerian models are more flexible when it comes to incorporating various physical and chemical processes (Eliassen 1984). It is
difficult to formulate general rules for when different types of models
should be used. The choice should depend on the spatial and temporal
coverage of measurement data, the availability of meteorological data,
etc.

Some examples are given below of different types of models that are
being used to simulate the long-range and the global transport of sulfur
and nitrogen compounds. Table 6-1 contains a list of models with a
further separation into spatial scales of the transport process with
brief remarks regarding their applicability.

6.2.3.1. Lagrangian Box Models. A well-mixed box of air extending from
the surface up to, for example, the top of the PBL, is followed along a
more or less schematic trajectory and transformation and removal processes are assumed to take place uniformly throughout the box. Dilution
with outside air may also be included. Examples of such models include
those of Rodhe et al. (1981, formation of sulfuric acid and nitric acid
through photochemical reactions during LRT), Eliassen et al. (1982,
formation of ozone during similar conditions) and Barrie and Hoff (1984,
formation of sulfate in polluted air moving to the Arctic).

6.2.3.2. Lagrangian Dispersion Models. These are similar to the Lagrangian box models but are being statistically applied to longer time
periods. To keep the computer time within bounds, chemical transformations are normally very simplified. The standard example is the EMEP
model used to simulate the dispersion of anthropogenic sulfur over Europe
(Eliassen and Saltbone 1983).

6.2.3.3. Horizontal Grid-point Models. These are employed to study the
distribution of sulfur compounds over Europe and North America (e.g.,
Johnson et al. 1978). Their application to remote areas where meteorological data are sparse is probably less meaningful at this stage.

6.2.3.4. One-dimensional Eulerian Models. These models can be useful
when studying the vertical distribution of gases and particulate matter

in the atmosphere when horizontal variations are small and most important
exchange processes take place vertically by exchange process on spatial
scales smaller than the scale characterizing the concentration differences under consideration. This may be the case, for instance, in the
stratosphere or in the atmospheric surface layer over the ocean or over
uniform land surfaces (Omstedt and Rodhe 1978). Hales (1982) has developed a comprehensive 1-D model with emphasis on cloud and scavenging
processes. These 1-D models have also been applied to study the exchange
between different latitude bands of, e.g., CO_2 (Bolin and Keeling 1963).

6.2.3.5. Two-dimension Eulerian Models. These models have been used to
study the global (height vs. latitude) distribution of nitrogen (Crutzen
and Gidel 1983) and sulfur compounds (Rodhe and Isaksen 1980). Some
results are shown in Figures 6-2 and 6-3. These exercises may be useful
in estimating hemospheric budgets and in checking the consistency of our
notions about transport, transformation, and removal processes. However,
the predicted concentration values should not be expected to agree
closely with all observations. In the PBL in particular, the limited
residence times of NO_x, HNO_3, SO_2, and sulfate particles imply that large
temporal and geographical variations occur. Because of the longitudinal
averages implicit in these kinds of model calculations, the models are
more suitable for species with residence times longer than the time for
zonal mixing around the globe (a few weeks). A major weakness of all the
previous applications of these 2-D models is their inability to account
for the rapid transport of boundary-layer air to the upper troposphere in
regions of substantial convective activity (Chatfield and Crutzen 1984).
In principle, such transport could be included in global 2-D models.

　　Two-dimensional grid-point models with height as one of the coordi-
nates may also be applied to transport simulations on smaller than global
scales. For example, Schutz (1979) has used such a model to simulate the
transport and deposition of Saharan dust across the Atlantic towards the
West Indies.

6.2.3.6. Three-dimensional Models. These general circulation models
(GCM) have been applied to global-scale studies of individual tracers,
including N_2O (Levy et al. 1982). Because of the computational work
involved, elaborate chemical or removal schemes can not yet be included
in these models. It is also doubtful whether the observational data base

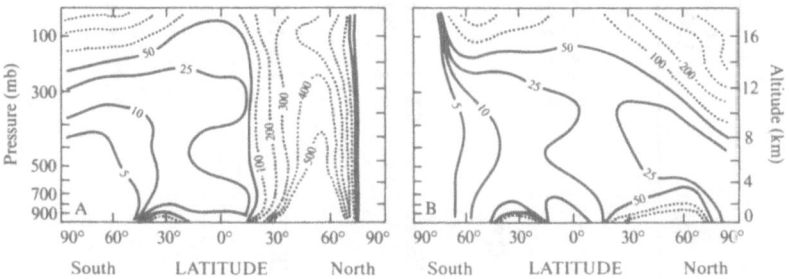

Figure 6-2. Calculated Meridional Distribution of NO_x Mixing Ratio in
　　　Precipitation. A: January, B: July (from Crutzen and Gidel 1983).

Table 6-1. Brief Overview of Transport Models.

Model Type	Horizontal Scale of Application		
	Local (<100 km)	Regional (100-300 km)	Global
		Eulerian	
1-D (V)*	OK for horizontally homogeneous situations (some boundary layers).		OK for stratosphere but not generally for troposphere.
1-D (H)*			Has been applied to N/S exchange of long-lived gases.
2-D (H/V)	OK for some situations with lateral homogeneity, e.g., line sources and steady cross winds.		Commonly used but severely limited in troposphere, particularly for short-lived compounds (DMS, SO_2, NO_x).
2-D (H/H)	Of limited use.	Used in some regional air-pollution studies, less useful in remote areas because of lack of data.	Of limited use.
3-D	Used in some urban diffusion studies. Less useful in remote locations because of lack of data.	Being developed for regional air-pollution studies.	Useful but costly to run; mainly for stable compounds (COS, N_2O, etc.); cloud and boundary-layer processes very poorly parameterized.

Lagrangian

For individual days	Sometimes used with trajectories, depending on local wind velocity.	Requires good wind data and limited vertical exchange, trajectories not accurate for more than 2-3 days.	Not applicable because of uncertainty of trajectories.
Average over season or longer	OK in principle.	Commonly used in regional air-pollution studies; trajectory time scales may be extended to several days.	Not applicable because of uncertainty of trajectories.

Other Types

Several exist	Example: Staubsauger model for deep mixing in convective regions (see text).	E.g., models for studies of S/N exchange

*1-D denotes one spatial dimension and (V) that this dimension is the vertical or (H), the horizontal.

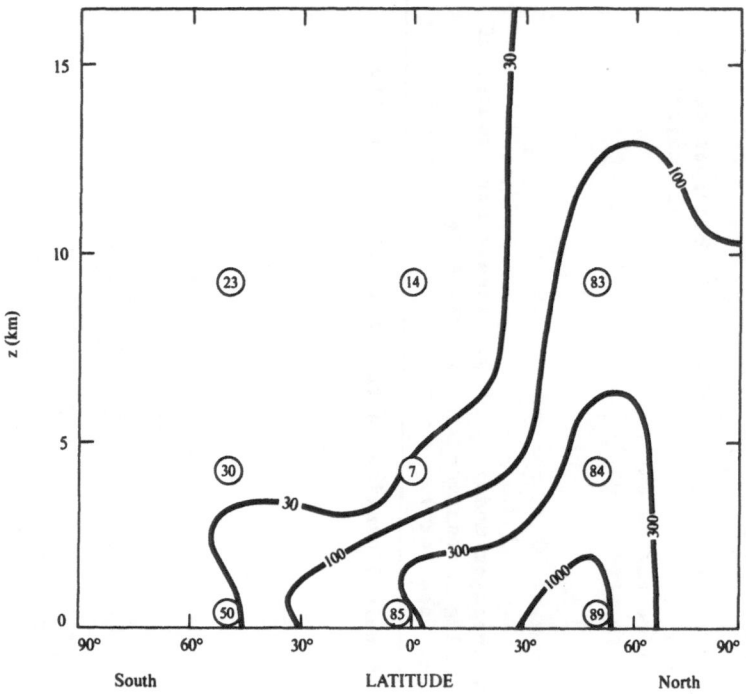

Figure 6-3. Calculated Distribution of $SO_4^=$ for July Resulting From a
        Release of 40 Tg H_2S (S)/yr and 80 Tg SO_2 (S)/yr (pptv). The
        numbers in circles give the percentage contributions of manmade
        sources to the local concentrations of $SO_4^=$ at the different heights
        and latitudes (from Rodhe and Isaksen 1980).

is good enough to verify a full 3-D simulation of sulfur or nitrogen
compounds.

    Of the several models that cannot be included in the above classi-
fication, the Staubsauger model of Chatfield and Crutzen (1984) should be
mentioned. They have studied the vertical transport of reactive sulfur
gases from the tropical marine PBL to the upper troposphere caused by
penetrative cumulonimbus clouds. They conclude that this mode of transport constitutes a very important mechanism for the transfer of certain
species from the PBL to the free troposphere. In regions with strong
convective activity and for species with a longer chemical lifetime in
the upper troposphere than in the lower (e.g., SO_2), this transport
mechanism may even create an inverse gradient with higher values in the
upper troposphere than in the middle layers (cf. Figure 6-4, see also
Gidel 1983). Such a transport process cannot be modeled either as a
normal advection process or as a diffusive process in a grid-point model
but requires a special parameterization.

6.3. CASE STUDIES

In this section, I briefly review situations where sulfur or nitrogen
compounds or both have been measured at remote sites and where the
possible origin of these compounds has been investigated using tracers or
long-range transport models.

6.3.1. Arctic Haze

Sulfate (at average levels of 1-2 $\mu g/m^3$) constitutes a dominant part of
the aerosol causing Arctic haze, i.e., an extensive haze covering large
parts of the Arctic region particularly during winter and spring months

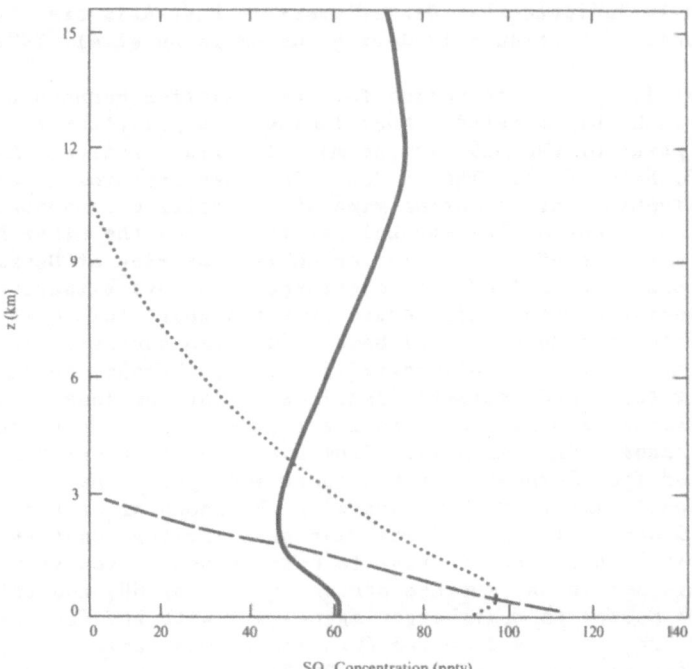

Figure 6-4. Characteristic Differences Between Eddy-diffusion Transport
and Cloud-transport Parameterizations. Concentrations of SO_2 are
shown as a function of height produced by emissions of dimethylsulfide and hydrogen sulfide as simulated by various authors. Model A
of Chatfield and Crutzen (1984, solid line) and that of Logan et al.
(1979, dotted line) simulate the emission of 0.067 $g(S)/(m^2 \cdot yr)$ of
dimethylsulfide. The work of Rodhe and Isaksen (1980, dashed line)
simulates production of SO_2 from 0.11 $g(S)/m^2 \cdot yr$) but the results
are scaled proportional to the sulfur emissions described above.
(Figure from Chatfield and Crutzen 1984).

(Rahn and Heidam 1981). The chemical and physical properties of the
Arctic haze in surface air have only recently been measured (cf. Proceedings of the Second Symposium on Arctic Air Chemistry, 1981 and Geophysical Research Letters, 1984, Vol. 11, No. 5) but haze layers at elevations
between 1 km and 6 km were already being observed in the late 1940's
(Raatz 1984). Simultaneous measurements of sulfate, several metals
(e.g., V, Mn, ^{210}Pb), and, at a few sites, SO_2 enable some inferences to
be made regarding possible sources of Arctic sulfate. Rahn (1981) has
reviewed the evidence regarding these sources.

     Arctic haze in surface air, including its sulfate content, has a
substantially anthropogenic origin. This conclusion is based mainly on
the high concentration of pollution-derived species, such as V, Mn, and
soot. Natural fractions derived from sea salt and from crustal material
have also been identified (Heidam 1981). Analyses of aircraft measurements of haze layers above the boundary layer in Alaska and the Canadian
Arctic strongly indicate that desert areas in East Asia can also be
important sources, particularly during summer (Rahn et al. 1977, Raatz
1984).

     The most likely source region for the pollution component of the
Arctic haze is highly debated. Most indications point towards Europe and
neighboring parts of the USSR as the most important sources (Larssen and
Hanssen 1980, Rahn 1981). The evidence includes meteorological considerations, geographical distribution maps of the sulfate concentration, and
chemical compositions of the aerosol particles. On the other hand, a
statistical analysis of receptor-oriented trajectories at Barrow, Alaska,
shows that more than half of the long-range-transport situations at that
site are associated with trajectories from the south (cf. Figs. 6-5 and
6-6). Patterson and Husar (1981) have calculated source-oriented 850-
mbar trajectories from three industrial regions (Europe, North America,
and Southeast Asia) and estimate significant contributions to the load of
sulfate at Barrow from all three source regions: from North America
mainly from January through April, from Southeast Asia between May and
September, and from Europe between October and April. However, the
median transport time from the sources to the receptor at Barrow ranges
from 13 to 19 days. In view of what was said earlier about the accuracy
of trajectories, these results have to be regarded as tentative.

     The decreases in the average concentrations of SO_2 and sulfate going
from northern Norway to Spitzbergen agree well with the decrease in the
frequency of trajectories from the Eurasian source region arriving at
these sites (Heintzenberg and Larssen 1983). Rahn et al. (1982) and
Barrie and Hoff (1984) have used simple transport models that include
chemical transformations to simulate episodes of sulfur transport from
the midlatitude source regions to the Arctic. By and large, these calculations reconcile the sulfate levels found in the Arctic with known or
estimated transformation and deposition rates. The levels of sulfate
found in the Arctic are also consistent with the indications of a longer
residence time of sulfur pollutants in Europe in winter than in summer
(Rodhe and Granat 1983).

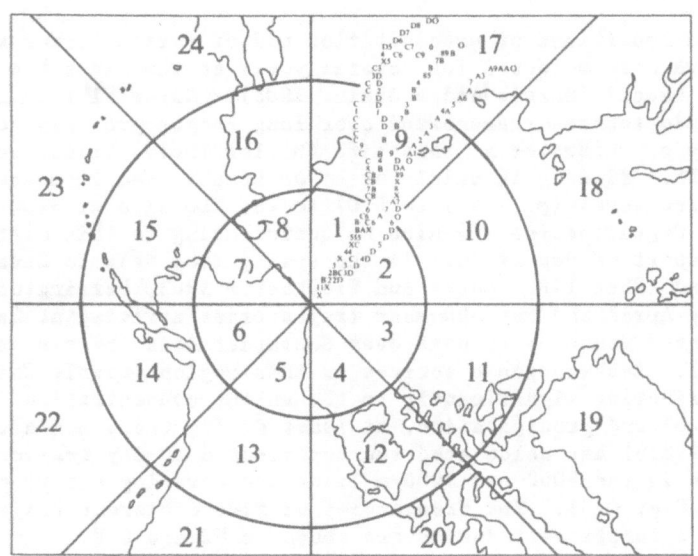

Figure 6-5. Classification of Barrow Trajectories With an Example of Trajectories from 12 August 1976. (Figure from Miller 1981a.)

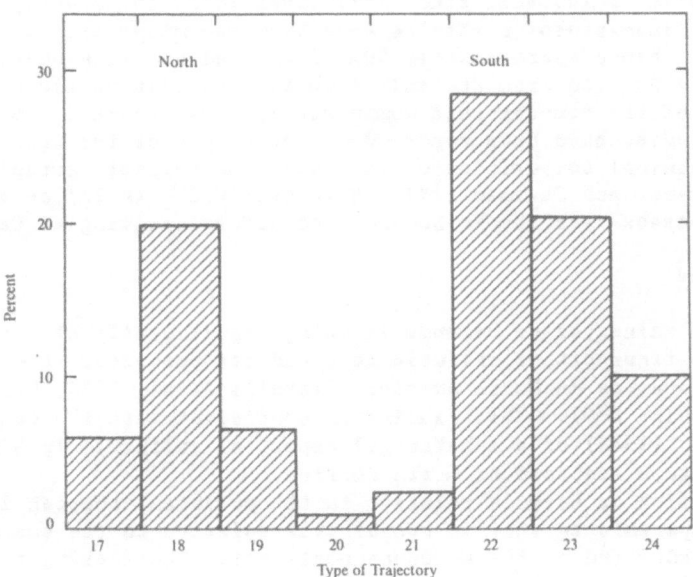

Figure 6-6. The Percentage of a Given Trajectory Type for a 5-yr Period. (Figure from Miller 1981a.)

6.3.2. Mauna Loa

The chemical compositions of precipitation and of various trace materials
in the air have been measured for several years at the Mauna Loa Observatory (MLO) in Hawaii (Harris and Bodhaine 1983). Material from natural
or anthropogenic sources transported over long ranges probably contribute
to the acidity of rainwater at this site (Miller 1981b, Miller and Yoshinaga 1981). This finding is mainly based on an observed increase in
acidity from sea level (pH = 4.7 to 5.0) to the MLO site at 3400 m (pH =
4.3 to 4.5). Organic acids may also be contributing to this difference.
     The transport of desert dust in the spring from Asia to Hawaii has
been documented (Shaw 1980, Darzi and Winchester 1982, Parrington et al.
1983). During April and May 500-mbar trajectories arriving at Hawaii
show a pronounced tendency to pass over Southeast Asia (Harris and
Bodhaine 1983). Anthropogenic sources in this region (mainly China) are
probably contributing significantly to the sulfur concentration in
Hawaiian aerosol and precipitation, at least during these months.
     Miller (1981b) has calculated source-oriented 10-day trajectories
based on winds in the 3000- to 5000-m layer and covering a 5-yr period at
the MLO site (Fig. 6-7). The frequencies of five different trajectory
directions as a function of season are shown in Figure 6-8. The overall
pattern is clearly defined: Westerly flows dominate in October through
May and easterly flows, in June through September.

6.3.3. Cape Grim

Precipitation-chemistry data have been available from the base-line station at Cape Grim in Tasmania since 1977 (Dyer 1983) and monthly values
of sulfate in fine-aerosol particles have been recently reported (Heintzenberg 1984). Since approximately 90% of the sulfate in precipitation
at this site is derived from sea salt (40% for the fine particles), the
determination of the non-sea-salt component is very uncertain (Dyer
1983). Few studies have been reported on the origin of the species affecting the chemical composition of rainwater and aerosol particles at
Cape Grim. Fraser and Pearman (1978) have used $CFCL_3$ (F-11) as a tracer
for air that passes over the Melbourne area before arriving at Cape Grim.

6.3.4. Bermuda

The acidity of rainwater on Bermuda is being significantly affected by
the long-range transport of sulfuric acid and its precursor (SO_2) from
anthropogenic sources in North America (Jickells et al. 1982, Church et
al. 1982; cf. Fig. 6-9). This finding is consistent with the estimate by
Galloway et al. (1984) of a substantial export of sulfur (4 Tg S/yr)
across the eastern seaboard of North America.
     The situation on Bermuda relative to the pollutant sources in North
America is comparable to that in Scandinavia relative to the sources in
the United Kingdom and on the European continent. Considering that the
geographical distances are similar, that the European sulfur emissions
are almost twice as large as those in North America, and that the frequency of airflow from the source regions to these receptors is somewhat

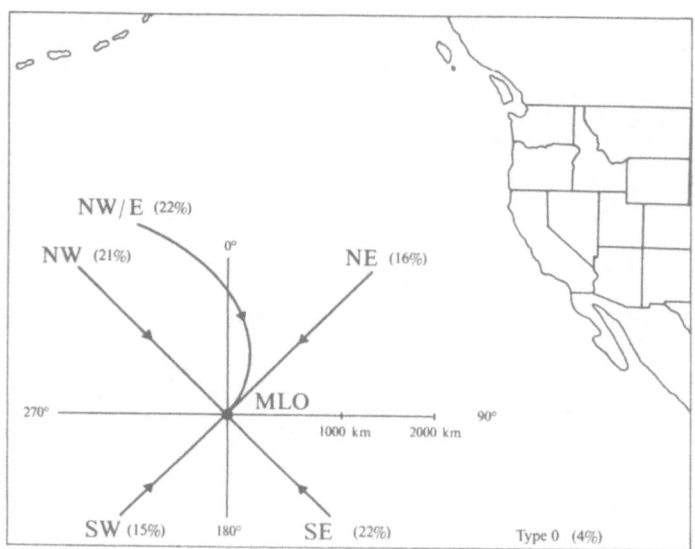

Figure 6-7. Trajectory-typing Method Used to Categorize MLO Trajectories. The percentage of a given trajectory type during the 5-yr period is also plotted. (Figure from Miller 1981b.)

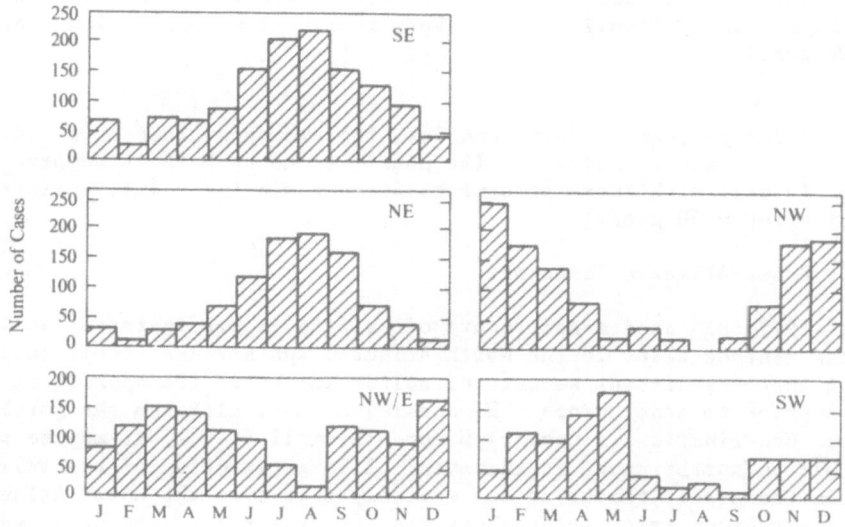

Figure 6-8. The Number of Trajectories of a Given Type Plotted for Each Month Over a 5-yr Period. (Figure from Miller 1981b.)

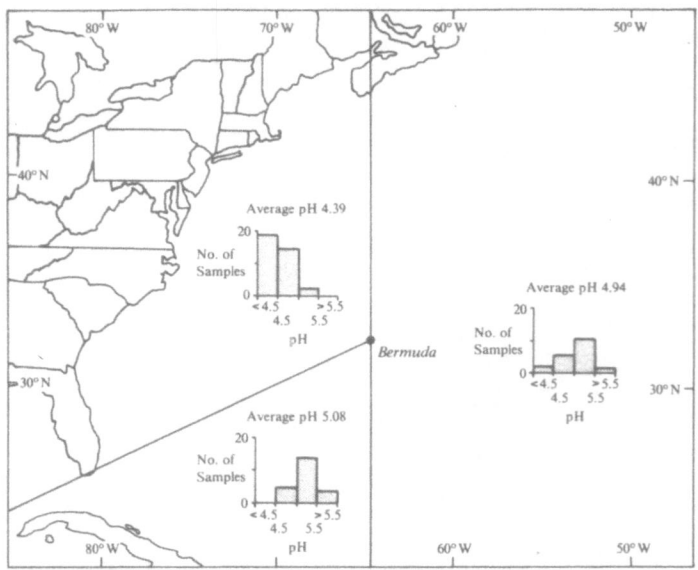

Figure 6-9. The Northwest Atlantic Ocean With the Three Sectors Used for
    ARL Back Trajectories of Bermuda Rain. The average pH of rain
    originating in each sector is indicated and the distribution of pH
    values for rain samples from that sector are shown in histograms.
    (Figure from Jickells et al. 1982, reprinted by permission from
    Nature.)

larger in Europe than in North America, the average values of pH (4.7)
and excess sulfate in rainwater (18 µeq/1) found on Bermuda compare
reasonably well with those of central Scandinavia (pH ~ 4.4, sulfate
concentration ~ 60 µeq/1).

6.3.5. Cross-Atlantic Transport

Nyberg (1977) has used measurements of sulfate in precipitation collected
on ocean weather ships in the North Atlantic and air-mass trajectories to
suggest that significant amounts of sulfur are being transported from
North America towards Europe. He concludes that, although the North
American contribution in Central Europe is small (<10%), it may be considerable in northernmost Scandinavia. A large fraction of the sulfate
found in the precipitation in the westerly flow over the North Atlantic
is probably being transported above the PBL from the source area and thus
being protected from dry deposition at the sea surface. This idea is
supported by the estimate of the export of sulfur from North America
presented by Galloway et al. (1984).

6.4. ACKNOWLEDGMENTS

This review has been improved by the constructive criticism of members of
the Transport Group at the Bermuda Workshop: L. A. Barrie, I. S. A.
Isaksen, J. M. Miller, F. B. Smith, and D. M. Whelpdale.

6.5. REFERENCES

Andreae, M. O, T. W. Andreae, R. J. Ferek, and H. Raemdonck. 1984. Long-
    range transport of soot carbon in the marine atmosphere. The
    Science of the Total Environment 36:73–80.

Barrie, L. A., and R. M. Hoff. 1984. Oxidation rate and residence time of
    SO_2 in the Arctic atmosphere. Atmos. Environ. 18:2711–2722.

Bolin, B., and C. D. Keeling. 1963. Large-scale mixing as deduced from
    seasonal and meridional variations of carbon dioxide. J. Geophys.
    Res. 68:3899–3920.

Chatfield, R. B., and P. J. Crutzen. 1984. Sulfur dioxide in remote
    oceanic air: Cloud transport of reactive precursors. J. Geophys.
    Res. 89:7111–7132.

Church, T. M., J. N. Galloway, T. D. Jickells, and A. H. Knap. 1982. The
    chemistry of western Atlantic precipitation at the mid-Atlantic
    coast and on Bermuda. J. Geophys. Res. 87:11013–11016.

Crutzen, P. J., and L. T. Gidel.'1983. A two-dimensional photochemical
    model of the atmosphere 2: The tropospheric budgets of the anthropo-
    genic chlorocarbons, CO, CH_4, CH_3Cl and the effect of various NO_x
    sources on tropospheric ozone. J. Geophys. Res. 88:6641–6661.

Danielson, E. F. 1961. Trajectories: Isobaric, isentropic, and actual.
    J. Meteor. 18:479–486.

Darzi, M., and J. W. Winchester. 1982. Aerosol characteristics at Mauna
    Loa Observatory, Hawaii, after east Asian dust storm episodes. J.
    Geophys. Res. 87:1251–1256.

Durst, C. S., A. F. Crossley, and N. E. Davis. 1959. Horizontal diffusion
    in the atmosphere as determined by geostropic trajectories. J. Fluid
    Mech. 6:401–422.

Dyer, A. J. (ed.). 1983. Baseline Air Monitoring Report 1979–1980.
    Canberra:Australian Govern. (Dept. of Sci. and Tech.).

Eliassen, A. 1984. Aspect of Lagrangian air pollution modeling. In
    Air Pollution Modeling and Its Application, III (C. Dewispelaere,
    ed.) New York:Plenum.

Eliassen, A., and J. Saltbone. 1983. Modeling of long-range transport of
    sulphur over Europe: A two-year model run and some model experi-
    ments. Atmos. Environ. 17:1457–1466.

Eliassen, A., O. Hov, I. Isaksen, J. Saltbone, and F. Stordal. 1982. A
    Lagrangian long-range transport model with atmospheric boundary
    layer chemistry. J. Appl. Meteor. 21:1645–1661.

Eriksson, E. 1959. The yearly circulation of chloride and sulfur in
    natureµ meteorological, geochemical and pedological implications:
    Part 1. Tellus 11:375–403.

Fraser, P. J., and G. I. Pearman. 1978. Atmospheric halocarbons--The
    CSIRO Southern Hemisphere programme. In Clean Air, The Continuing
    Challenge (E. T. White, P. Hetherington, B. R. Thiele, eds.), Ann
    Arbor, MI:Ann Arbor Science, 703–716.

Galloway, J. N., and G. E. Likens. 1978. The collection of precipitation for chemical analysis. Tellus 30:71-82.

Galloway, J. N., D. M. Whelpdale, and G. T. Wolff. 1984. The flux of S and N eastward from North America. Atmos. Environ. 18:2595-2607.

Gidel, L. T. 1983. Cumulus cloud transport of transient tracers. J. Geophys. Res. 88:6587-6599.

Grey, D. C., and M. L. Jensen. 1972. Bacteriogenic sulfur in air pollution. Science 177:1099-1100.

Hales, J. 1982. Mechanistic analysis of precipitation scavenging using a one-dimensional, time-variant model. Atmos. Environ. 16:1775-1783.

Harris, J. M., and B. A. Bodhaine. 1983. Geophysical Monitoring for Climatic Change No. 11: Summary Rept. 1982. Washington, DC: U. S. Govern. Printing Office (NOAA 84041801), 174 pp.

Heidam, N. Z. 1981. On the origin of the Arctic aerosol: A statistical approach. Atmos. Environ. 15:1421-1427.

Heintzenberg, J. 1984. Physical and chemical aerosol characteristics in extremely clean air masses at Cape Grim, Tasmania. Presented at the 11th Int. Conf. on Atmos. Aerosols, Budapest, Sept. 1984. (Paper available from Dr. Heintzenberg, Dept. of Meteorology, Univ. of Stockholm.)

Heintzenberg, J., and S. Larssen. 1983. SO_2 and SO in the Arctic: Interpretation of observations at three Norwegian Arctic-sub-Arctic sites. Tellus 35B:255-265.

Hitchcock, D. R., and M. S. Black. 1974. $^{34}S/^{32}S$ evidence of biogenic sulfur oxides in a salt marsh atmosphere. Atmos. Environ. 18:1-17.

Jickells, T., A. Knap, T. Church, J. Galloway, and J. Miller. 1982. Acid rain on Bermuda. Nature 297:55-57.

Johnson, W. B., D. E. Wolf, and R. L. Manenso. 1978. Long-term regional patterns and transformation exchanges of airborne sulphur pollution in Europe. Atmos. Environ. 12:511-527.

Junge, C. E. 1963. Air Chemistry and Radioactivity. New York:Academic Press.

Krouse, H. R. 1980. Sulfur isotopes in our environment. In Handbook of Environmental Isotope Geochemistry, Vol. 1: The Terrestrial Environment (P. Fritz and J. U. Fontes, eds.), Amsterdam:Elsevier, 435-471.

Larssen, S., and J. E. Hanssen. 1980. Annual variations and origin of aerosol components in the Norwegian Arctic/sub-Arctic region. In WMO Tech. Conf. on Regional and Global Observations of Atmos. Pollut. Relative to Climate (WMO Publication #549), Geneva, Switzerland:WMO, 251-258.

Levy, I. I., J. D. Mahlman, and W. J. Moxim. 1982. Tropospheric N_2O variability. J. Geophys. Res. 87:3061-3080.

Logan, J. A., M. B. McElroy, S. C. Wofsy, and M. J. Prather. 1979. Oxidation of CS_2 and COS: Source for atmospheric SO_2. Nature 281:185-186.

Miller, J. M., 1981a. A five-year climatology of back trajectories from the Mauna Loa Observatory, Hawaii. Atmos. Environ. 15:1553-1558.

Miller, J. M., 1981b. A five-year climatology of five-day back trajectories from Barrow, Alaska. Atmos. Environ. 15 1401-1405.

Miller, J. M., and J. M. Harris. 1985. The flow climatology to Bermuda and its implications for long-range transport. (Manuscript available from Dr. Miller, NOAA-ARL, Silver Spring, MD.)

Miller, J. M., and A. M. Yoshinaga. 1981. The pH of Hawaiian precipitation: A preliminary report. Geophys. Res. Letters 8:779-782.

Mizutani, Y., and T. A. Rafter. 1969. Isotopic composition of sulfate in seawater, Gracefield, New Zealand. N.Z.J. Sci. 12:69-80.

Nielsen, H. 1974. Isotopic composition of the major contributions to atmospheric sulfur. Tellus 26:213-221.

Nriagu, J. O., and R. D. Coker. 1978. Isotopic composition of sulfur in precipitation within the Great Lakes Basin. Tellus 30:365-375.

Nyberg, A. 1977. On airborne transport of sulphur over the North Atlantic. Quart. J. R. Met. Soc. 103:607-615.

Omstedt, G., and H. Rodhe. 1978. Transformation and removal processes for sulfur compounds in the atmosphere as described by a one-dimensional time-dependent diffusion model. Atmos. Environ. 12: 503-509.

Ostlund, G., 1959. Isotopic composition of sulfur in precipitation and seawater. Tellus 11:478-480.

Pack, D. M., J. E. Lovelock, G. Cotton, and C. Curthoys. 1977. Halocarbon behavior from a long time series. Atmos. Environ. 11: 329-344.

Pack, D. H., G. J. Ferber, J. L. Heffter, K. Telegadas, J. K. Angell, W. H. Hoecker, and L. Machta. 1978. Meteorology of long-range transport. Atmos. Environ. 12:425-444.

Parrington, J. R., W. H. Zoller, and N. K. Aras. 1983. Asian dust: Seasonal transport to the Hawaiian Islands. Science 220:195-197.

Patterson, D. E., and R. B. Husar. 1981. A direct simulation of hemispherical transport of pollutants. Atmos. Environ. 15:1479-1482.

Raatz, W. E. 1984. Observations of "Arctic Haze" during the "Ptarmigan" weather reconnaissance flight, 1948-1961. Tellus 36B:126-136.

Rahn, K. A. 1981. Relative importance of North America and Eurasia as sources of Arctic aerosol. Atmos. Environ. 15:1447-1455.

Rahn, K. A., and N. Z. Heidam. 1981. Progress in Arctic air chemistry, 1977-1980: A comparison of the first and second symposia. Atmos. Environ. 15:1345-1348.

Rahn, K. A., and D. H. Lowenthal. 1984. Elemental tracers of distant regional pollution aerosols. Science 223:132-139.

Rahn, K. A., R. D. Borys, and G. E. Shaw. 1977. The Asian source of Arctic haze bands. Nature 265:713-715.

Rahn, K. A., C. Brosset, B. Ottar, and E. M. Patterson. 1980. Black and white episodes, chemical evolution of Eurasian air masses and long-range transport of carbon to the Arctic. In Particulate Carbon, Atmospheric Life Cycle (G. T. Wolff and R. L. Klimirsch, eds.), New York:Plenum.

Rodhe, H. and L. Granat. 1983. Summer and winter budgets for sulfur over Europe: An indication of large seasonal variations of residence time. Idojaras 87:1-6.

Rodhe, H., and I. Isaksen. 1980. Global distribution of sulfur compounds in the troposphere estimated in a height/latitude transport model. J. Geophys. Res. 85:7401-7409.

Rodhe, H., C. Persson, O. Akesson. 1972. An investigation into regional transport of soot and sulfate aerosols. Atmos. Environ. 6:675-693.

Rodhe, H., P. Crutzen, and A. Vanderpol. 1981. Formation of sulfuric and nitric acid in the atmosphere during long-range transport. Tellus 33:132-141.

Schutz, L. 1979. Saharan dust transport over the North Atlantic Ocean--
    model calculations and measurements. In SCOPE 14:Saharan Dust (C.
    Morales, ed.), New York:Wiley.
Seiler, W., and J. Fishman. 1981. The distribution of carbon monoxide and
    ozone in the free troposphere. J. Geophys. Res. 86:7255-7265.
Shannon, I. D., and D. E. Patterson. 1983. Continental and hemispheric
    transport. In Acidic Deposition Phenomenon and Its Effects. Critical
    Assessment Review Paper, Vol. 1: Atmospheric Sciences. Washington,
    DC:U. S. Govern. Printing Office (EPA).
Shaw, G. E. 1980. Transport of Asian desert aerosol to the Hawaiian
    Islands. J. Appl. Meteor. 19:1254-1259.
Sisterson, D. L., J. D. Shannon, and J. M. Hales. 1979. An examination of
    regional pollutant structure in the lower troposphere: Some results
    of the diagnostic atmospheric cross section experiment (DACSE-1).
    J. Appl. Meteorol. 19:1421-1428.
Smith, J. W. 1975. Stable isotope studies and biological element cycl-
    ing. In Environmental Chemistry (G. Eglington, ed.), London:Chemi-
    cal Society, 1-21.
Thode, H. G., and C. E. Rees. 1970. Sulphur isotope geochemistry and
    Middle East oil studies. Endeavor XXIX:24-28.

7. THE TRANSPORT OF SULFUR AND NITROGEN THROUGH THE REMOTE ATMOSPHERE WORKING GROUP REPORT

John M. Miller
Air Resources Laboratory
NOAA
8060 13th Street
Silver Spring, MD 20910

Douglas M. Whelpdale
Environment Canada
Atmospheric Environment Service
4905 Dufferin Street
Downsview, Ontario, Canada

Leonard A. Barrie
Environment Canada
Atmospheric Environment Service
4905 Dufferin Street
Downsview, Ontario, Canada

Ivar S. A. Isaksen
Box 1022
Institute of Geophysics
University of Oslo
Blindern, Oslo 3, Norway

Henning Rodhe
Department of Meteorology
Arrhenius Laboratory
University of Stockholm
S-10691 Stockholm, Sweden

F. Barry Smith
British Meteorological Office (Met. O. 14)
London Road
Bracknell RG 12 2SZ, England

7.1. INTRODUCTION

The transport of sulfur and nitrogen species in the troposphere takes place on many different spatial scales, depending on where the species are released and how efficient the transformation and removal processes are, both en route and in the particular area of interest. A better

J. N. Galloway et al. (eds.), The Biogeochemical Cycling of Sulfur and Nitrogen in the Remote Atmosphere, 127–139.
© 1985 by D. Reidel Publishing Company.

knowledge of these processes is highly desirable. Sulfur and nitrogen species are released in localized areas through man's activities on the earth's surface and through natural processes, e.g., in tropical rain forests, over the oceans, in agricultural areas, and even in the free troposphere where nitrogen oxides can be formed by lightning (see Chapter 2). Such meteorological phenomena as clouds, precipitation, and wind speed play decisive roles in the processes determining the transport, transformation, and removal of most sulfur and nitrogen species from different regions of the troposphere (cf. Chapter 5).

To emphasize the global variability of the meteorological factors important in sulfur and nitrogen cycling, during our working-group sessions in Bermuda we divided the troposphere into a limited number of regimes (see Fig. 7-1 and Table 7-1). Each regime is characterized by reasonably uniform meteorological processes. This is rather a crude description and is intended to serve only as a general framework for assessing sulfur and nitrogen cycles in the troposphere. The boundaries in Figure 7-1 are not fixed but should be moved with the seasons.

As shown schematically in Figure 7-1, we have divided the troposphere into seven geographical regions. Four of these regions are within the tropospheric boundary layer (polar, midlatitude, subtropical, and equatorial) and three in the free troposphere (polar night, midlatitude, and tropical). Because seasonal variations are large at midlatitudes and high latitudes, particularly over the continents, a distinction has to be made between summer and winter regimes at these latitudes. Furthermore, the planetary boundary layer has very different characteristics over land than it does over oceans. Distinguishing between land and sea surfaces is also important in estimating the release or the deposition of sulfur and nitrogen species.

Whether sulfur and nitrogen species are transformed and removed within a geographical regime or transported to another regime depends on their lifetimes. For sulfate and nitrate particles (as well as gaseous HNO_3), precipitation scavenging represents a major removal mechanism. If the time scale for removal is shorter than the time scale for mixing with other regions, most of what is emitted in or formed within a specific region will be deposited in that region.

In the gas phase, SO_2 is converted in the atmosphere mainly by the reaction of OH and, in the liquid phase, most likely by reaction with O_3 and H_2O_2 (see Chapter 5). Consequently, gas-phase chemistry and cloud and precipitation processes play significant roles in the availability of these substances in the atmosphere. For many of the other sulfur and nitrogen species, gas-phase oxidation—in most cases, oxidation by OH—is the major oxidation mechanism in the troposphere. Chemical lifetimes may vary from hours or days (e.g., NO_2 oxidation to HNO_3 and DMS oxidation of SO_2), which leads to efficient oxidation within most geographical regimes, to months or years (e.g., COS oxidation to SO_2), which allows an almost complete mixing within the whole troposphere. Obviously, realistic estimates of chemical lifetimes are crucial in determining the distances that S and N species are being transported, especially in the free troposphere.

There are also large variations in the cloud and precipitation parameters **within** a given region and these variations were not reflected

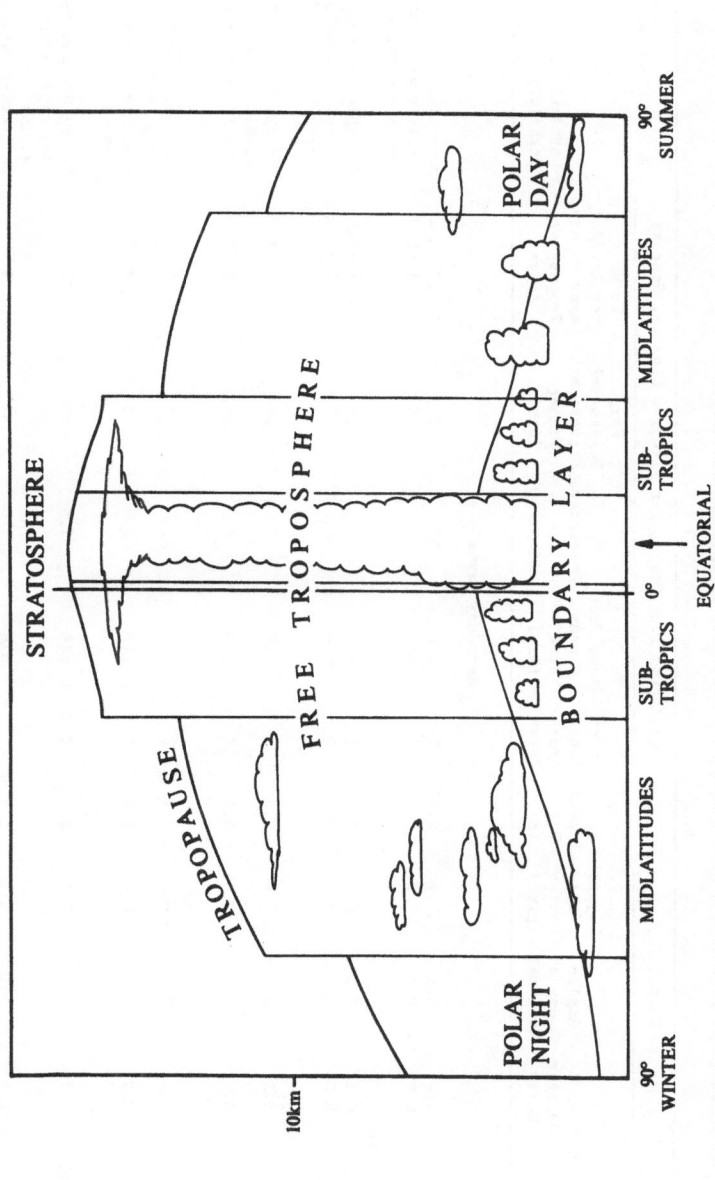

Figure 7-1. Meteorological Regimes. These regimes usually correspond to specific land types and ecological regions in response to distinctive precipitation characteristics even though the correspon-
dence cannot be complete because of the latitudinal shift of the regimes with seasons. The amount
and type of vegetation within each zone, over land, can be classified as follows if, for simplicity,
the moving regimes are roughly linked with fixed latitudinal bands: 60°-90° = 40% tundra, 60%
grassland, 7% snow; 30°-60° = 30% arable farmland, 40% grassland, 30% forest; 10°-30° = either 50%
grassland or 25% desert, 35% grassland, and 30% forest (see also Table 7-1).

Table 7-1. Meteorological Regimes and Some Preliminary Estimates of Their Characteristics.

| | Cloud | | | | Estimated Time Scales for Airborne Chemical Species | | | | | |
	Average Cover	Region Occupied (volume)	Liquid Water Content (g/m³)	Annual Ppt (m)	To Enter Cloudy Area* (days)	To Enter Cloud from Cloudy Area** (hrs)	To Encounter Ppt (days)	For Horizontal North-South Mixing (days)	To Move From Boundary Layer to Free Troposphere (days)	Vertical Stability
					Boundary Layer					
Polar										
Winter	0.4	0.3	<0.1	0.05	2	≤10	20	15	>20	Very stable
Summer	0.6	0.4	0.2	0.15	1	≤10	5	15	5 (?)	Stable over oceans; often neutral over land
Midlatitude: Winter										
Continental	0.4	0.2	0.1	0.2	2	≤10	2	25	5 (?)	Variable; neutral, stable
Oceanic	0.3	0.2	0.3	0.7	1	≤10	2	25	3 (?)	Variable; usually near-neutral
Midlatitude: Summer										
Continental	0.2	<0.1	0.4	0.3	1	≤10	4	20	5 (?)	Unstable: neutral by day; stable at night
Oceanic	0.4	0.1	0.4	0.3	1	≤10	4	20	5 (?)	Usually near neutral
Subtropical										
Continental	<0.1	0.1	0.2	0.1	3	≤10	15	5	2 (?)	Very unstable by day; deep mixing
Oceanic	0.2	<0.1	0.4	0.5	3	≤10	8	5	4 (?)	Near neutral; inversion-capped
Equatorial										
Continental	0.3	<0.1	0.8	2	0.5	≤10	1.5	≤2	<2	Strong diurnal cycle; by day rapid tropospheric mixing
Oceanic	0.3	<0.1	0.8	1.5	1	≤10	2	≤2	<2	No diurnal cycle; rapid tropospheric mixing

					Free Troposphere				
Polar night	<0.1 (?)	<0.1	0.1	n/a	?	?	?	n/a	Gradual subsidence; stable
Midlatitude	0.4	0.1	0.3	n/a	2	?	?	n/a	Sporadic mixing by cloud
Tropical	0.1	<0.1	1.0	n/a	3	?	?	n/a	Hadley cell circulation; rapid mixing in equatorial zone
				Stratosphere					
	0	0	-	0	-	-	-	n/a	Stable

Sources: Values for average cloud cover and volume of occupied region are based on Warren et al. (1983); low clouds (Cu, St, Sc) are assumed to be located within the boundary layer. Values for liquid content of clouds are based on Hegg (1985); liquid-water content varies significantly with cloud type; values correspond to dominant cloud type wherever possible. Time to encounter precipitation is based on Hamrud (1984); values are very uncertain.

Note: Values marked with ? are particularly uncertain and require upgrading by field-experiment data.

*Order of magnitude only; assume emission in cloud-free area.

**A single value, based on simple considerations of boundary-layer depth and typical vertical velocities, is given for all regions in the absence of available experimental information. In a situation with a complete cover of low clouds (e.g., Sc), this time scale may be as short as half an hour.

Table 7-2. Preliminary Estimates of Oxidant Concentrations and Average Residence Times of Key S and N Species in Different Meteorological Regimes.

	Oxidant Concentration[†]			Average Residence Times[††]						
	OH molec/cm³ (×10⁵)	Mixing Ratio		SO₂[a,c,d] (hrs)	DMS[d] (hrs)	NOₓ (NO+NO₂) (hrs)	HNO₃[b,c] (hrs)	PAN	NO₃/N₂O₅*	Submicron Aerosol Particles (hrs)
		H₂O₂ (ppbv)	O₃ (ppbv)							
				Boundary Layer						
Polar										
Winter	<1	<0.2	20-40	>100**	>250**	≤30ᶜ	<10	slow**	?	500**
Summer	3-10	0.4-2	15-30	≤50	~30	30-70ᶜ,ᵈ	~30	fast	fast	120
Midlatitude: Winter										
Continental	(1-10)	(0.1-2)	20-40	<50**	>50	~30-120**ᶜ,ᵈ,ᵉ	≤120	hrs-days	fast	50**
Oceanic	1-4	0.2-2	20-40	≤50	>150	50-150ᵈ,ᵉ	≤20	hrs-days	---	50
Midlatitude: Summer										
Continental	(3-15)	(0.2-5)	20-40	≤30	~20	≤30**ᶜ,ᵈ	≤40	fast	---	100**
Oceanic	2-5	1-5	20-40	≤50	~100**	50-120**ᶜ	≤30	fast	---	100**
Subtropical										
Continental	(3-15)	(0.2-2)	15-30	≤50	~20	20-50ᶜ,ᵈ	≤100	fast	---	350**
Oceanic	2-5	1-5	15-30	≤10	~50	50-120ᶜ	≤50	fast	---	200**
Equatorial										
Continental	(2-15)	(0.1-2)	15-40	≤50**	~20	20-70**ᶜ,ᵈ	≤25	fast	---	25-50
Oceanic	2-5	1-5	15-30	≤10	~50**	50-120**ᶜ	≤40	fast	---	50**

Free Tropospere

Polar night	<1	<0.2	20-60	many	>250	≤50	>100**	weeks**	slow**(?)	many**
Midlatitude Winter	3	0.1-2	20-60	≤50	25-100	≤100**	>100**	weeks	<30**	many **
Summer	12	0.1-2	10-30	≤50	~30	≤100**	?	weeks	<30**	?
Tropical	12	0.1-2	10-30	≤50	~30	≤100**	?	weeks	<30**	?

Notes: The letters a–d refer to the different process that are taken into account as important when the residence time for the different species is estimated: a = incloud scavenging of SO_2 by H_2O_2 and O_3, b = precipitation scavenging, c = dry deposition, d = gas-phase reaction with OH, e = gas-phase reactions converting NO_2 to NO_3/N_2O_5 or PAN.

†Two-dimensional model estimates of OH and H_2O_2 distributions are uncertain, particularly in continental areas exposed to large releases of gaseous and particulate matter, in such cases numbers are given in parentheses. Values of O_3 are observed values.

††DMS reaction rate with OH is $k = 9 \cdot 10^{-12}$ cm^3/(molec · s); HNO_3 deposition effective on all surfaces ($v_d \approx$ 1 cm/s); PAN determined by thermal decomposition in the boundary layer, strongly temperature dependent, reaction with OH may be important in the free troposphere.

*Dissociated efficiently to give NO_2 back when sunlight is present.

**Regimes where the exchange with other regimes may be sufficiently rapid to affect the residence time. An important transport out of the region may take place.

in the average numbers in Table 7-1. The framework, however, should be
useful for determining a general picture of the transport of S and N
species on scales smaller than the global scale.

7.2. RESIDENCE TIMES

When attempting to estimate regional budgets, one must have some knowledge about the average residence times of the various compounds in the
atmosphere. This time scale is a measure of how long it would take, on
the average, for a molecule introduced into the atmosphere to be transformed to some other compound or to be deposited on the ground. If the
residence time within a certain portion of the atmosphere is considered,
the transport out of that portion would also influence the residence
time. In different situations, the average residence time is limited by
the different transformation and removal processes (see Chapters 5 and
9): Gas-phase transformation rates are governed by the concentration of
oxidants in the air--for several S and N compounds, the most important
one is OH. However, for liquid-phase transformation, particularly in
ocean areas where the concentration of NO_x is low, the rate-limiting
process may be the transport from cloud-free air to within a cloud.

   Using the meteorological data in Table 7-1, in particular those
referring to clouds and precipitation, in conjunction with estimated and
measured concentrations of the most important oxidants, we have attempted
to derive rough estimates of the average residence times within the
various regions of some key species discussed at the workshop in Bermuda
(Table 7-2). Table 7-2 points out the large variations in average residence time between different meteorological regimes. However, the actual
numbers in this table are very uncertain and should only be used in other
contexts with great care. There is an almost complete lack of reliable
measurements of the concentrations of the oxidants OH and H_2O_2 in different parts of the atmosphere. The estimates of the concentrations in the
first two columns of Table 7-2 are based mainly on calculations from two-
dimensional photochemical models (Rodhe and Isaksen 1980).

7.3. FUTURE FIELD AND MODELING STUDIES

Field measurements to improve our understanding of the processes related
to atmospheric chemical cycles are urgently needed. Those involving
transport are generally expensive and require aircraft support, costly
sophisticated measurement and computational equipment, and the cooperation of several research groups. Experimental design and interpretation
should be done using theoretical models as guides. Measurements alone
are useful but their value would be greatly enhanced by being set into a
theoretical framework.

   We have divided proposed field and modeling studies in which transport would be an important feature into two groups. The first group
should be designed to acquire an improved understanding of the transport-
exchange processes between different regimes in the atmosphere (e.g., the
regimes depicted in Fig. 7-1). The second should be designed to clarify
particular aspects of transformation and removal processes (in rather
simple meteorological regimes). Such regimes should exhibit properties

of steadiness or horizontal homogeneity that would reduce the uncertainties associated with physical transport. Consequently, estimates of transformation and deposition would be more reliable.

7.3.1. Transport-exchange Studies

7.3.1.1. Vertical Mixing by Convective Clouds.

Convective cloud fields effect a significant redistribution of pollutants in the vertical that cannot be described by simple eddy-diffusion theory. Experiments are needed to develop a more sophisticated treatment that would account for the dynamics of the convective cloud fields (i.e., by providing evaluation data sets for theory).

Two situations were especially mentioned at the workshop in Bermuda. The first is the enhanced exchange between midlatitude tropical and equatorial boundary layers and the free troposphere associated with fair-weather cumulus fields. One possible experiment would involve releasing an inert tracer (e.g., perfluorocarbon) from a line source for several hours while measuring downwind vertical profiles throughout an atmospheric layer containing a cloud field.

The second situation was the mixing of the equatorial troposphere by deep cumulonimbus cloud fields. For instance, theoretical models of convective cloud fields have indicated that substances emitted at the earth's surface can be rapidly delivered to the upper troposphere in updrafts within a cloud (Chatfield and Crutzen 1984). To study this process using tracers would require considerably more effort than to examine fair-weather cumulus mixing--a much larger volume of air would be involved that would require sampling through a greater range of heights. In this situation, natural tracers might provide better alternatives than artificial ones would.

7.3.1.2. Atmospheric Pathways Through the Upper Equatorial Troposphere.

In the equatorial midtroposphere, the oxidation of atmospheric nitrogen to NO_2 by lightning and the subsequent photochemical reactions produce significant amounts of nitrate (Logan 1983). Some of this nitrate may be finding its way to polar ice fields on Greenland (Herron 1982). Existing global circulation models should be used to study the possible pathways for this transport.

7.3.1.3. Dry Pathways.

In the previous examples, clouds are an important component in the exchange processes between regimes (particularly in transporting boundary-layer pollutants into the free troposphere). However, important exchanges almost certainly arise from the spatial and temporal character of the underlying terrain in which clouds are not playing an important role. For instance, mountain ranges can sometimes create a strong disturbance of tropospheric flow causing boundary-layer pollutants to be stirred through a much deeper layer of the atmosphere before the boundary layer is regenerated further downwind. Another example is the successive diurnal development of the boundary layer over land, which, in response to changes in surface conditions, may fail to re-entrain pollution left aloft on earlier days. Onshore and offshore winds provide an important example of this latter phenomenon. Although the first process could be studied using aircraft-determined profiles of

natural or specifically introduced tracers upwind and downwind of such
mountain ranges, the second process would be much harder to study experimentally. However, the extent of its importance could be assessed by
considering mixing depths obtained from numerical models incorporating
radiosonde data.

7.3.1.4. <u>Polar-night Boundary Layers</u>. A few scientific groups are
stationed in the Antarctic all year. (see Chapter 8). In spite of the
extremely difficult working conditions, these groups study the behavior
of pollutants in the polar-night boundary layer, which is often very
stable and is rather shallow over the extensive snow fields. At 2 m or
3 m above ground, the wind speed may be very low but, at 10 m or 20 m,
it is usually quite high. Such strong shears can become dynamically
unstable and result in sporadic, short-lived bursts of turbulence causing
the downward transport of pollutants. Studies investigating this phenomenon should be mounted to establish whether or not it is an important
sink for some compounds.

7.3.2. Simple Transport Regimes

7.3.2.1. <u>Light-wind Slow-moving High-pressure Systems</u>. Many parts of
the world experience, at least sometimes, large anticyclonic systems that
are very slow moving and have low associated winds. These can occur over
either land or sea during either winter or summer. If the pollutant to
be studied has rather uniform natural sources and sinks in the area
(e.g., oceanic areas, tundra, forests, etc.), transformation and deposition processes could be studied in a one-dimensional (vertical) frame.
For instance, studies of a subtropical Atlantic high pressure over the
Sargasso Sea might yield information on the natural emission of sulfur
compounds.

7.3.2.2. <u>Steady-wind Situations over Long Fetches of Homogeneous
Terrain</u>. In a steady-wind situation over a long stretch of homogeneous
terrain, the vertical profile of a pollutant might enable assumptions to
be made about the horizontal constancy of emissions, vertical mixing, and
transformation. Any one of these three factors could be estimated by
measuring the other two. For example, in a steady trade-wind regime free
of cumulus clouds (i.e., below 1500 m), the transformation of DMS might
be obtained from profiles by inferring the profile of eddy diffusivity
and the emission rate of DMS. Another such situation would be the flow
regime in the roaring forties of the Southern Hemisphere.

7.3.2.3. <u>Monsoons</u>. In atmospheric-chemistry studies, a meteorological
condition that we felt might prove useful was the monsoon. In some parts
of the world, seasonal variations in temperature differences between
continents and oceans produce a corresponding variation in winds. Such
well-organized seasonal patterns are called monsoons, of which the south-
western flow into India during October through June is probably the most
well known. The northern Australian monsoon has been investigated for
its precipitation-chemistry properties by Galloway et al. (1982). Because a monsoon pattern has well-defined seasonal boundaries (November to

March), this study could help delineate natural sources of chemical species.

7.3.2.4. Pollutant Transformation in Fair-weather Cumuli. At midlatitudes over continents under conditions of broad-scale subsidence, fair-weather cumulus fields can develop below an inversion at 1.5-2 km. During their growth and evaporation cycle of 30 min to 1 hr, these clouds subject boundary-layer air to a mixed-phase environment of air and cloud drops with diameters of 1-50 μm. The rate of the processing of air by clouds may vary widely depending on cloud cover, depth, and updraft velocity (cf. Table 7-1). During this process, soluble sulfur and nitrogen compounds can undergo chemical and physical transformations. In this relatively simple situation, it might be possible to infer transformation rates by following an air mass in a Lagrangian sense with aircraft while measuring the vertical profiles of aerosol composition and gas concentration before and after the cloud field has undergone its daytime cycle.

7.3.2.5. Orographic Effects: Precipitating Clouds. Orographically induced precipitation has the obvious advantage of providing a quasi-stationary system in which uptake, transformation, and loss of incoming, natural (as well as man-made) pollutants could be studied at fixed points. This could be achieved without the extremely difficult task of following the development relative to a moving Lagrangian parcel in moving precipitation bands. Obvious examples of such areas that could be largely unaffected by local anthropogenic sources are the Norwegian Mountains in northwest Europe and the Hawaiian Islands in the Pacific.

7.3.2.6. Orographic Effects: Nonprecipitating Clouds. Hilly areas frequently have more clouds than do lowland areas. Many of these clouds form in the boundary-layer air as it is forced over the hills. The pollutants carried in the boundary-layer air are subject to in-cloud transformation processes that could be meaningfully studied at a series of fixed monitoring sites. An interesting example of this is the wave clouds that sometimes develops above and downwind of mountain ranges (Hegg and Hobbs 1982).

7.3.3. Measurement Strategies from a Meteorological Perspective

7.3.3.1. The Global Approach. Following the meteorological regimes we have outlined in this chapter, an overall monitoring strategy could be developed. Such a strategy should be designed to determine both the mean concentrations and variances of the compounds of interest. Nonmeteorological considerations, such as the location of different ecosystems, could also be incorporated into the design (see Chapter 3). Thus, taking into account the meteorological and ecological factors, an idealized monitoring network could be designed simply by placing in each regime or ecosystem one or several measuring sites. Measurements should at least include the important sulfur and nitrogen species recommended in Parts I, II, and III of this report ($SO^2_4{}^-$, $NO_3{}^-$, $NH_4{}^+$ in wet deposition and aerosols and SO_2, DMS, NO, NO_2, HNO_3, and NH_3 in air).

In view of the approximate homogeneity of meteorological conditions and thereby, to a certain extent, the homogeneity of vegetation types

within each regime, variations in time-averaged concentrations within
each region may be smaller, or at least not larger, than possible systematic variations between the regimes. If this is true, a reasonable
estimate of global deposition could be obtained from a network encompassing only 10 or 20 stations. On the other hand, if we are to allow
for the possibility of significant localized sources within the regimes
and if the global-flux estimate is to be highly precise, many more than
10 measurement stations would be required within each regime, increasing
to a few hundred for a global network.

7.4. SUMMARY AND RECOMMENDATIONS

At the time of the Bermuda workshop, past research had shown that a
simple meteorological evaluation of remote atmospheric-chemistry data
could in many cases be performed using back-trajectory analyses (Chapter 6). Whereas the uncertainty from using estimated individual trajectories is very substantial, using many trajectories might give reasonable
information about the origin of the air masses, several days before,
affecting a particular site. We agreed that the use of tracers, anthropogenic as well as natural, is another potentially powerful method of
identifying the nature and the location of sources of a measured species.
However, the application of detailed atmospheric models would only be
valid if a sufficiently large meteorological and chemistry data base were
available.
    Some attempts have been made to quantify the long-range transport of
S and N species (e.g., industrial regions to certain remote locations).
However, these attempts are generally limited by a lack of quantitative
knowledge about transformation and removal processes as well as of
transport processes.
    A framework of meteorological regimes was developed during the
Bermuda workshop that is tailored to the chemical cycles of S and N
compounds in the atmosphere-earth system. We hope that this framework
will be used by scientists investigating emission and transformation
questions as a tool to organize their study of S and N cycles.
    During our group discussions, several simple meteorological sce-
narios with reasonably well-known meteorological parameters were defined
and recommended for further study. In this way, emissions, transformations, and deposition could be studied without unnecessary complications.
    As a final step our group combined the framework of meteorological
regimes with other factors, such as, ecological considerations. A
logical approach to establishing regional chemical budgets of sulfur and
nitrogen compounds would require that research into the transport of
these species be integrated with other aspects of atmospheric chemistry.

7.5. REFERENCES

Chatfield, R. B., and P. J. Crutzen. 1984. Sulfur dioxide in remote
    oceanic air: C transport of reactive precursors. J. Geophys. Res.
    89:7111-7132.
Galloway, J. N., G. E. Likens, W. C. Keene, and J. M. Miller. 1982. The
    composition of precipitation in remote areas of the world. J.
    Geophys. Res. 87:8771-8786.

Hamrud, M. 1984. Lagrangian time scales connected with clouds and precipitation. Rept. CM-65, Department of Meteorology, University of Stockholm.

Hegg, D. A. 1985. The importance of liquid-phase oxidation of SO_2 in the troposphere. J. Geophys. Res. (in press).

Hegg, D. A., and P. V. Hobbs. 1982. Measurements of sulfate production in natural clouds. Atmos. Environ. 16:2663-2668.

Herron, M. M. 1982. Impurity sources of F^-, Cl^-, NO_3^- and $SO^2_4^-$ in Greenland and Antarctic precipitation. J. Geophys. Res. 87:3052-3060.

Logan, J. A. 1983. Nitrogen oxides in the troposphere: Global and regional budgets. J. Geophys. Res. 88:10,785-10,807.

Rodhe, H., and I. Isaksen. 1980. Global distribution of sulfur compounds in the troposphere estimate in a height/latitude transport model. J. Geophys. Res. 85:7401-7407.

Warren, S. G., C. Hahn, and J. London. 1983. Distribution of six cloud types over the oceans and their diurnal and interannual variations. In Preprint Vol: Fifth Conference on Atmospheric Radiation. Boston: Am. Meteorol. Soc.

PART IV

THE DEPOSITION OF SULFUR AND NITROGEN FROM THE REMOTE ATMOSPHERE

8. THE DEPOSITION OF SULFUR AND NITROGEN FROM THE REMOTE ATMOSPHERE BACKGROUND PAPER

James N. Galloway
Department of Environmental Sciences
University of Virginia
Charlottesville, VA 22903

8.1. INTRODUCTION

S and N species are deposited from the atmosphere by both wet and dry processes. The magnitudes of these processes have been estimated for over a century. As time goes by, past estimates are refined and estimates for species previously unknown are determined.

There are two major motivations for determining deposition rates and concentrations of S and N species in atmospheric deposition. First, measurements of atmospheric deposition determine the removal rates of materials from one reservoir and the input rates into another (e.g., terrestrial and aquatic ecosystems). These removal and input rates set boundary conditions on the biogeochemical processes within the reservoirs. Second, the measurement of concentrations in atmospheric deposition aids in determining the important biogeochemical reactions in the atmosphere. Simply put, the concentration at which something is found in atmospheric deposition helps to determine the major reaction pathways in the atmosphere and the lifetime of the constituent measured in the atmosphere.

In this chapter, I present estimates of the wet- and dry-deposition rates for selected S and N species to remote areas using recent publications on S and N deposition in remote areas and an assortment of unpublished data sets. At the end of each discussion of a specific species, I have chosen an estimate of the deposition rate for that species for comparison purposes only. These estimates have uncertainties varying from a factor of 2 to a factor of 100, which are caused by many different reasons. In the rest of this introduction, I discuss the sources of some of these uncertainties before discussing the actual data.

This chapter served as the beginning point of a discussion on deposition at the workshop and I thank the participants of the working group on deposition and Alex Pszenny and Dennis Savoie for their astute and helpful comments on it. In the following chapter, the participants of the working group on deposition present an assessment of the extent of the knowledge and a refinement of some of the material presented here. There is also an expanded discussion of the uncertainties in the estimates presented here.

J. N. Galloway et al. (eds.), The Biogeochemical Cycling of Sulfur and Nitrogen in the Remote Atmosphere, 143–175.
© 1985 by D. Reidel Publishing Company.

8.1.1. Wet Deposition

Because of the isolation of remote areas, two types of problems are
involved in determining wet-deposition rates for S and N species.   The
first problem concerns the quality of the data base.   Several aspects
need to be considered--the length of the sampling period, the sample-
collection techniques, the sensitivities and interferences of the analytical techniques used, the chemical changes during storage, and the errors
involved in estimating nonmeasured parameters (e.g., excess $SO_4^=$).   Programs that collect wet deposition on greater than an event or weekly
basis are generally only suitable for determining very rough deposition
rates for S and N species because of sample evaporation, contamination,
or actual chemical changes in the sample as it sits in the collector.
Relative to analytical sensitivity, wet-deposition samples from remote
continental areas typically have $SO_4^=$, NO_3^-, and NH_4^+ concentrations that
range from 0.5-10 μeq/l.   Thus, an analytical technique with an accuracy
and precision of at least ±0.1 μeq/l may be required.
     Wet-deposition samples from remote marine areas have similar analy-
tical limitations but with the additional problem of differentiating the
excess $SO_4^=$ from the sea-salt $SO_4^=$.   The determination of excess $SO_4^=$
depends on two analyses, one for the total $SO_4^=$ and one for the sea-salt
parameter (e.g., magnesium or sodium) that serves as a tracer for the
sea-salt $SO_4^=$.   Thus, the analytical errors possible from using these two
analyses and the problems associated with subtracting two large numbers
to obtain the excess $SO_4^=$ concentration incorporate the possibility of a
substantial margin of error (Keene et al. 1985, see also Chapter 9 for
additional discussions).
     Second, because the data bases are limited, any significant temporal
variations (e.g., winter-versus-summer differences) or spatial variations
(e.g., differences between marine-versus-continental or wet-versus-dry
areas) can obscure attempts to determine an accurate annual deposition
rate for either S or N species.
     In the following subsections on wet deposition, I present recent
estimates of the rates of deposition of S and N species with explanations
of how the estimates have been determined and the possible limitations of
the accuracy of these estimates.   Each subsection concludes with an
estimate of the wet-deposition rate for comparison to deposition rates
for other species of S and N.

8.1.2. Dry Deposition

Our understanding of the temporal and spatial variabilities of wet deposition is imperfect because of the lack of high-quality data available on
the actual rate of deposition.   For dry deposition, two additional reasons further limit our ability to determine deposition rates.
     First, the art of measuring actual dry-deposition rates on natural
surfaces is poorly developed.   As an alternative, surrogate surfaces are
used, however, there are no acceptable ways of validating the deposition
rates collected on these surfaces.
     Second, in the absence of measured dry-deposition rates, measure-
ments of atmospheric concentrations and empirically or theoretically

derived deposition velocities are used to calculate dry-deposition rates.
The errors associated with both techniques can be large. Not only are
the spatial and temporal variabilities of the atmospheric concentrations
poorly defined for remote areas but also instrument sensitivity can be
inadequate for several important species (e.g., NO_2, HNO_3, PAN). Published values of deposition velocities for individual S and N species
vary over several orders of magnitude. Selecting the appropriate value
for a given calculation requires a detailed knowledge of surface characteristics, several micrometeorological variables, and the chemical and
physical characteristics of the depositing material (Sehmel 1980). In
addition the concept of deposition velocity itself may not be valid for
some species (see Part I and Chapter 9).

     In the sections on dry deposition, I present recent estimates of
rates of dry deposition of S and N species with explanations of how the
estimates have been determined. These subsections conclude with an estimate of the dry-deposition rate and a comparison to deposition rates for
other species of S and N. Chapter 9 contains an additional discussion on
the phenomenon of and problems associated with dry deposition.

8.2. SULFUR

8.2.1. Introduction

This discussion of the deposition of S from the atmosphere to remote
areas considers SO_2, $(CH_3)_2S$, sea-salt $SO_4^=$, and excess $SO_4^=$. Three
types of programs are presently being used to determine wet-deposition
rates in remote areas: long-term programs that sample wet deposition on a
monthly basis (e.g., World Meteorological Organization [WMO]), short-term
programs that sample wet deposition on an event basis (e.g., Sea-Air
Exchange [SEAREX]), and long-term programs that sample wet deposition on
an event basis (e.g., Global Precipitation Chemistry Program [GPCP]).
Because of the problems discussed earlier, I have not used wet-deposition
data from programs with a monthly sampling period. My estimates of the
rates of dry deposition have been based on atmospheric concentrations and
deposition velocities.

8.2.2. Marine Areas: Wet Deposition

Several authors have used wet-deposition data, aerosol data, or both to
calculate the wet-deposition rate of excess $SO_4^=$ and sea-salt $SO_4^=$ from
remote marine atmospheres (Granat et al. 1976, Bonsang et al. 1980,
Andreae 1982, Mészáros 1982, Ryaboshapko 1983, Varhelyi and Gravenhorst
1983, Pszenny et al. 1982, Savoie 1984). Summaries of these $SO_4^=$ fluxes
(as S) show a range of 0.03-0.41 g S/(m^2 · yr) for excess $SO_4^=$ and 0.05-
0.6 g S/(m^2 · yr) for sea-salt $SO_4^=$ (Table 8-1). There are several
possible reasons for the large ranges--different methods of collection,
unrepresentative sampling periods, unrepresentative sampling locations,
sampling or analytical problems, or actual large-scale spatial variability in the wet-deposition rate of $SO_4^=$. The following discussion sum-
marizes the past estimates of wet deposition of excess and sea-salt $SO_4^=$,
compares these estimates to more recent data, and then presents a best
estimate of these deposition rates to remote marine areas.

Table 8-1. Wet Deposition of Excess and Sea-salt $SO_4^=$ to Remote Marine Areas.

Location	Number of Samples	Excess $SO_4^=$ Concentration (μeq/l)	Deposition (g S/(m²·yr))	Sea-salt $SO_4^=$ Concentration (μeq/l)	Deposition (g S/(m²·yr))	Reference
Global Estimates						
	–	20	0.03	–	–	Granat 1976
	–	–	0.41	16	0.3	Mészáros 1982
	–	–	0.09*	12±6	0.4±0.1	Ryaboshapko 1983
	–	–	0.15–0.2	6–56	0.3–0.5	Varhelyi & Gravenhorst 1983
	–	–	0.065	–	0.28	Savoie 1984
Southern Hemisphere				*Temperate Zone*		
	–	–	0.076	–	–	Bonsang et al. 1980
	–	–	0.074	–	–	Andreae 1982
	–	–	0.1	–	0.2–0.4	Varhelyi & Gravenhorst 1983
	–	–	0.095	–	–	Savoie 1984
			Pacific Ocean			
Samoa	12	2.0	0.066	9.7	0.3	Pszenny et al. 1982
New Zealand	8	3.3	0.067	13	0.3	Pszenny & Duce, unpubl.
Central Tasman Sea	4	6.3	0.13	23	0.5	Pszenny & Duce, unpubl.
			Indian Ocean			
Indian Ocean	–	–	0.1–0.3	–	0.3–0.6	Varhelyi & Gravenhorst 1983
Amsterdam Island	66	4.9	0.1	24.3	0.5	Galloway & Gaudry 1984
	–	–	0.05	–	–	Savoie 1984
			Atlantic Ocean			
North Atlantic 0°–30° N	–	–	0.2–0.3	–	0.1–0.2	Varhelyi & Gravenhorst 1983
	–	–	0.08	–	–	Savoie 1984
Western Atlantic						
Bermuda	19	9.5	0.23	10.8	0.26	Galloway, unpubl.
Ship†	27	6.1	0.14	16.2	0.39	Galloway et al. 1983; Galloway et al., unpubl.
South Atlantic	–	–	0.05–0.1	–	0.05–0.1	Varhelyi & Gravenhorst 1983
	–	–	0.12	–	–	Savoie 1984

*My estimate based on Ryaboshapko (1983, see text for details).

**From storms associated with air masses from the central Atlantic Ocean to minimize the influence from North America.

†See Figure 8-4 for tracks of ships.

8.2.2.1. **Excess Sulfate.** The largest value for wet deposition of excess $SO_4^=$ is 0.41 g S/(m^2 · yr) (Mészáros 1982), which is based on precipitation data from Pago Pago, Samoa. Varhelyi and Gravenhorst (1983) have estimated the excess $SO_4^=$ in wet deposition to remote marine areas by analyzing a compilation of wet-deposition-chemistry and aerosol-chemistry data. Using the wet-deposition-chemistry data, the calculated excess $SO_4^=$ fluxes in the Southern Hemisphere range from 0.01-2.8 g S/(m^2 · yr). Using data on excess $SO_4^=$ concentrations in aerosols and a $SO_4^=$ washout ratio of 1.5 · 10^{-6}, Varhelyi and Gravenhorst calculate a value of 0.03 g S/(m^2 · yr) for excess $SO_4^=$ in wet deposition. They then select appropriate weighting factors and calculate an average wet deposition of 0.15-0.2 g S/(m^2 · yr) to the world's oceans. Using subsets of this data base, they estimate wet-deposition rates for excess $SO_4^=$ of 0.1 g, 0.1-0.3 g, 0.2-0.3 g, and 0.05-0.1 g S/(m^2 · yr) for the Southern Hemisphere, Indian Ocean, North Atlantic Ocean (0°-30° N), and South Atlantic Ocean, respectively (Table 8-1).

Based on a review of seven papers published between 1957 and 1977, Ryaboshapko (1983) concludes that the excess $SO_4^=$ concentration in marine wet deposition is 12 ± 6 μeq/l. This, coupled with an annual precipitation rate of 128 cm/y, gives an annual rate of deposition of excess $SO_4^=$ of 0.25 g ± 0.12 g S/(m^2 · yr). This value, however, contains both a natural and an anthropogenic component. Although Ryaboshapko discusses the preindustrial S cycle, he only estimates a total $SO_4^=$ (excess plus sea salt) deposition rate and not an excess $SO_4^=$ deposition rate. Ryaboshapko does, however, estimate the emission of excess and sea-salt $SO_4^=$ to the marine atmosphere. Assuming that the two types of $SO_4^=$ are deposited in the same ratio as they are emitted, I deduced a wet-deposition rate of excess $SO_4^=$ to remote marine areas of 0.09 g S/(m^2 · yr) from the data in Ryaboshapko (1983) (Table 8-1).

Based on aerosol data collected in September 1979 at Cape Grim, Tasmania, and Townsville, Queensland, Andreae (1982) calculates a wet-deposition rate for excess $SO_4^=$ of 0.074 g S/(m^2 · yr). Bonsang et al. (1980) calculate wet-deposition rates for excess $SO_4^=$ of 0.076 g S/(m^2 · yr) from aerosol samples taken in the sub-Antarctic Indian Ocean and off the northwestern African coast.

Granat (1976), in an attempt to eliminate samples contaminated with anthropogenic $SO_4^=$, has selected only the lowest wet-deposition rates of $SO_4^=$ available and from these he calculates a rate of 0.03 g S/(m^2 · yr).

Pszenny and Duce (unpublished data available from A. A. P. Pszenny, Department of Environmental Sciences, University of Virginia, Charlottesville), as part of a SEAREX program, have calculated deposition rates of 0.067 g and 0.13 g S/(m^2 · yr) from samples collected on New Zealand (90-Mile Beach) and on board a ship in the Tasman Sea, respectively. Pszenny et al. (1982) report a similar value of 0.066 g S/(m^2 · yr) for their samples from Samoa.

Savoie (1984) has determined the geographical and temporal distribution of the concentration of excess $SO_4^=$ in the boundary layer from over 1,000 bulk aerosol samples. Deposition rates are calculated from these data by using a washout ratio of 0.3 m^3/cm^3. The resulting rates are combined with other published results to estimate an excess $SO_4^=$ flux of 0.065 g S/(m^2 · yr) by wet deposition to the world's oceans. Using subsets of this data base, Savoie (1984) estimates wet-deposition rates for excess $SO_4^=$ of 0.095 g, 0.05 g, 0.08 g, and 0.12 g $S/(m^2 \cdot yr)$ for the Southern Hemisphere, Indian Ocean, Atlantic Ocean (0°-30° N), and the South Atlantic Ocean, respectively.

As part of the Global Precipitation Chemistry Project, Galloway and Gaudry (1984) estimate that the wet-deposition rate of excess $SO_4^=$ at Amsterdam Island in the Indian Ocean is 0.1 g $S/(m^2 \cdot yr)$, based on 30 months of event sampling.

Two studies (unpublished data available from the author, Galloway et al. 1983) on wet deposition of excess $SO_4^=$ to the Atlantic Ocean give values of 0.23 g and 0.14 g $S/(m^2 \cdot yr)$, respectively (Table 8-1). These deposition rates are calculated using an annual precipitation rate of 150 cm.

The estimates of excess $SO_4^=$ in wet deposition discussed above range from 0.03 g to 0.41 g $S/(m^2 \cdot yr)$. I felt the value of 0.41 g $S/(m^2 \cdot yr)$ (Mészáros 1982) was too high for the following reasons. It is based on data from only one site--Pago Pago, Samoa. The average excess $SO_4^=$ concentration estimated for this site (20 $\mu eq/l$) is significantly greater than for other remote marine areas (Table 8-1). The Pago Pago data could be too high because precipitation is collected on a monthly basis or because of an error in estimating the element concentration used as a sea-salt tracer. Sampling periods of precipitation longer than a week can cause serious problems in data quality (Galloway and Likens 1976, 1978, Cogbill et al. 1984). These possibilities are discussed by Logan (1983a) and Mészáros (1983).

At the other extreme of the range, the value of 0.03 g $S/(m^2 \cdot yr)$ (Granat 1976) is less than any other estimate in Table 8-1 and probably provides a lower limit for excess $SO_4^=$ deposition in remote marine areas.

With the elimination of the 0.41 g $S/(m^2 \cdot yr)$ value, the revised range is thus 0.03-0.3 g $S/(m^2 \cdot yr)$ (Table 8-1). The arithmetic mean and standard deviation of the 20 values in this range are 0.11 ± 0.06. It is remarkable that these 20 values agree so well as some are based on a few measurements of rain or aerosol samples and others are based on large data sets of rain and aerosol analyses. The remaining variability is most assuredly caused by regional differences in the composition of atmospheric aerosols, in the amount of precipitation, and, perhaps, in a regional difference in the washout rate (Savoie 1984).

Based on the information above, wet-deposition rates for excess $SO_4^=$ in remote marine areas appear to be on the order of 0.1 g $S/(m^2 \cdot yr)$ (Fig. 8-1). This value needs to be validated by establishing additional sites in remote marine areas to collect precipitation on an event basis over at least one year.

8.2.2.2. <u>Sea-salt Sulfate</u>. Estimates of the deposition of sea-salt $SO_4^=$ in remote marines areas range from 0.05 g to 0.6 g $S/(m^2 \cdot yr)$ (Table 8-1). These fluxes are determined in the same way as excess $SO_4^=$ in wet deposition is--from precipitation measurements and atmospheric aerosol concentrations. Varhelyi and Gravenhorst (1983) have estimated a range of wet deposition of sea-salt $SO_4^=$ of 0.3-0.5 g $S/(m^2 \cdot yr)$ for the global oceans. Using subsets of these data, they also estimate wet-deposition rates of 0.2-0.4 g, 0.3-0.6 g, 0.1-0.2 g, and 0.05-0.1 g $S/(m^2 \cdot yr)$ for the Southern Hemisphere oceans, Indian Ocean, Atlantic Ocean

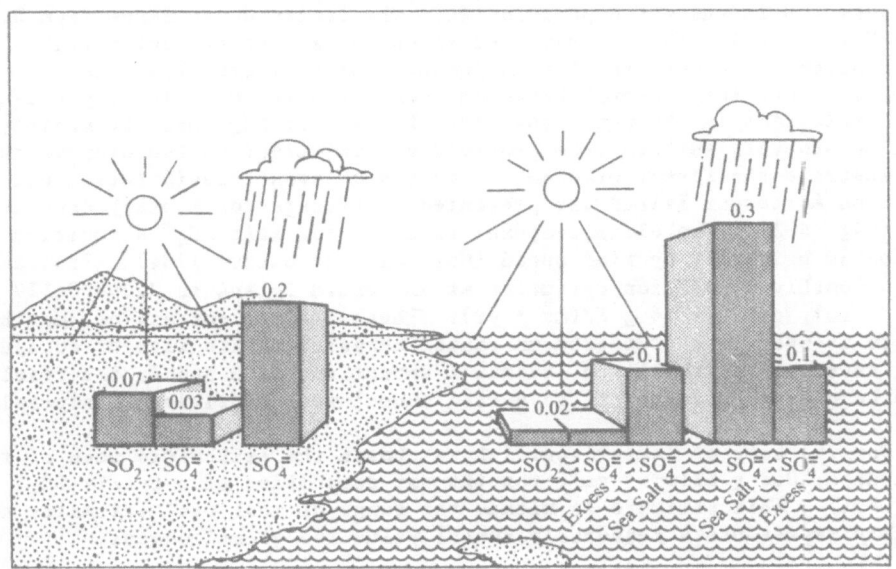

Figure 8-1. Wet- and Dry-deposition Rates of S Species in Remote Areas (g S/[m^2 · yr]).

(0°-30° N), and the South Atlantic Ocean, respectively. For the global ocean, Mészáros' (1982) estimate is 0.3 g S/(m^2 · yr) and Ryaboshapko's (1983) estimate is 0.4 g S/(m^2 · yr). Galloway and Gaudry (1984), using 30 months of data from Amsterdam Island, estimate a wet-deposition rate for sea-salt $SO_4^=$ in the Indian Ocean of 0.5 g S/(m^2 · yr). Two studies from the Western Atlantic Ocean estimate that the wet-deposition rate of sea-salt S is 0.26 g and 0.39 g S/(m^2 · yr) (Table 8-1). Pszenny et al. (1982) and Pszenny and Duce (unpublished data available from A. A. P. Pszenny, Department of Environmental Sciences, University of Virginia, Charlottesville), using the same data as discussed previously for excess $SO_4^=$, estimate sea-salt $SO_4^=$ wet deposition to be 0.3 g, 0.3 g, and 0.5 g S/(m^2 · yr) for Samoa, New Zealand, and the Central Tasman Sea, respectively. Savoie (1984) uses an extensive data base on aerosol composition and a washout ratio of 1.0 SCM/cm^3 to estimate a global ocean wet-deposition rate of sea-salt $SO_4^=$ of 0.28 g S/(m^2 · yr) (Table 8-1).

The range of the 14 referenced studies for sea-salt-$SO_4^=$ deposition covers an order of magnitude, 0.05-0.6 g S/(m^2 · yr). However, since most studies fall within a much narrower range, there is a small standard deviation about the arithmetic mean (0.3 g $\pm$ 0.1 g S/[m^2 · yr]). As in the case of excess $SO_4^=$, the small degree of deviation is suprising given the variety of sampling programs from which the estimates in Table 8-1 were taken. The amount of variability is again caused by many different factors. Among these are spatial and temporal variabilities in precipitation amount and perhaps the washout ratio. As an example of the magnitude of some of these problems, I have analyzed the data from the Indian Ocean.

As summarized in Varhelyi and Gravenhorst (1983), individual values of sea-salt S in the wet deposition from the Indian Ocean range from 0.1–1.8 g S/(m^2 · yr). The programs for which these data are collected sample either by event (short time periods) or by month (long time periods). Short-term records based on event sampling may miss any seasonal variations in the wet deposition of sea-salt $SO_4^=$ and the monthly sampling probably suffers from problems of contamination and evaporation. To illustrate the former problem, 30 months of event precipitation collected on Amsterdam Island are presented as averages of monthly deposition (Fig. 8-2). The clear temporal trend in sea-salt $SO_4^=$ deposition is controlled primarily by wind speed (Galloway and Gaudry 1984). The range of the monthly deposition estimates at Amsterdam Island is 0.008–0.120 g S/(m^2 · mo), or 0.1–1.4 g S/(m^2 · yr). Thus, if data for any one month were to be used to extrapolate an annual deposition rate without considering seasonal variability of rainfall amount and wind speed, the result would be a poor estimate of the actual annual wet deposition of sea-salt $SO_4^=$.

For the sake of comparison, I have chosen the arithmetic mean of the 14 studies, 0.3 g S/(m^2 · yr), for the wet deposition of sea-salt $SO_4^=$ (Fig. 8-1). This estimate is within the range of most past estimates and probably represents a reasonable annual value for global oceans.

8.2.3. Marine Areas: Dry Deposition

8.2.3.1. Sulfur Dioxide. Recent estimates of dry-deposition rates of SO_2 in remote marine areas range from 0.01–0.03 g S/(m^2 · yr)

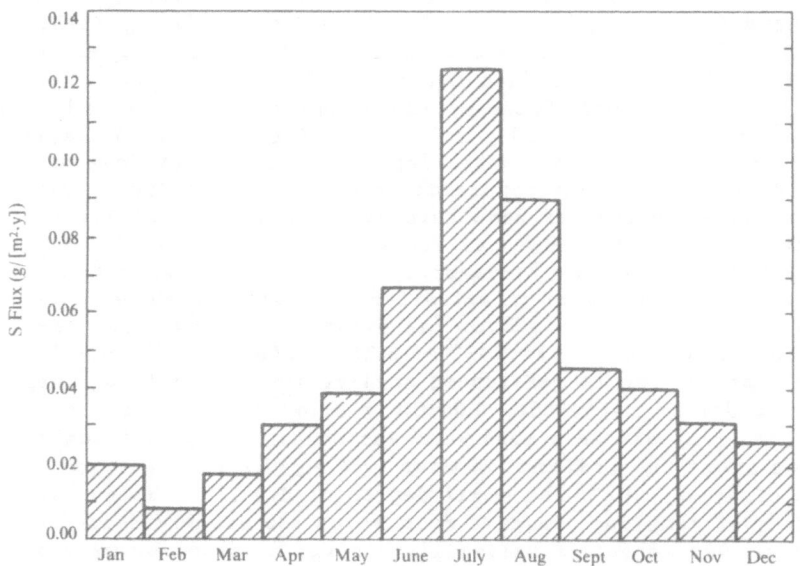

Figure 8-2. The Monthly Volume-weighted Mean Sea-salt $SO_4^=$ Concentration in Wet Deposition on Amsterdam Island (µeq/l).

Table 8-2. Concentrations, Deposition Velocities, and Dry-deposition
            Fluxes for Sulfur Species in Remote Marine Areas.

Atmospheric Concentration (μg S/m^3)	Deposition Velocity (cm/sec)	Dry-Deposition Rate (g S/[m^2 · yr])	Reference
\multicolumn{4}{c}{Sea-salt SO$_4$$^=$}			
0.02–0.4	1.5±0.5	0.07–0.20	Varhelyi & Gravenhorst 1983
0.15±0.16	0.5	0.02	Ryaboshapko 1983
0.3	–	0.12	Mészáros 1982
\multicolumn{4}{c}{Excess SO$_4$$^=$}			
0.02–0.05	1.5±0.5	–	Varhelyi & Gravenhorst 1983*
0.1–0.2	0.05±0.03	0.006–0.033	Varhelyi & Gravenhorst 1983**
0.35±0.15	0.2	0.02	Ryaboshapko 1983
0.3	–	0.02	Mészáros 1982
0.22	0.05	0.011	Savoie 1984
\multicolumn{4}{c}{SO$_2$}			
0.05–0.1	0.7	0.01–0.02	Varhelyi & Gravenhorst 1983
0.1	0.7	0.02	Mészáros 1982
0.1–0.2	0.8±0.2	0.03	Ryaboshapko 1983
0.05	0.5	0.01	Granat et al. 1976

*r > 0.5 μm.
**r < 0.5 μm.

(Table 8-2). Reasonable values for SO$_2$ concentrations and deposition
velocities in remote marine areas are 0.1 μg S/m^3 and 0.7 cm/sec, respectively. For purposes of this discussion, I chose the value of 0.02 g
S/(m^2 · yr) for the dry deposition of SO$_2$ to remote marine areas (Fig. 8-
1). However, this choice is based more on a lack of data than any complete knowledge.

8.2.3.2. _Sulfate_. Ryaboshapko (1983), using an atmospheric concentration
of 0.35 ± 0.15 μg S/m^3 and a dry-deposition velocity of 0.2 cm/sec,
estimates that the dry deposition of excess SO$_4$$^=$ is 0.02 g S/(m^2 · yr).
He also estimates the same rate for sea-salt SO$_4$$^=$, using a lower SO$_4$$^=$
concentration (0.15 μg ± 0.16 μg S/m^3) and a higher deposition velocity
(0.5 cm/sec) because of the larger particle size (Table 8-2).
    Varhelyi and Gravenhorst (1983) have estimated that the rates of dry
deposition of excess and sea-salt SO$_4$$^=$ are 0.006–0.033 g S/(m^2 · yr) and
0.07–0.20 g S/(m^2 · yr), respectively. For excess SO$_4$$^=$, they use the
concentration of 0.1–0.2 μg S/m^3 for oceanic areas other than the North

Atlantic and a deposition velocity of 0.05 ± 0.03 cm/sec for 75% of the excess $SO_4^=$ and of 1.5 ± 0.5 cm/sec for the remaining 25% of the excess $SO_4^=$ mass. The size distribution of excess $SO_4^=$ over the oceans suggests such a partition. For sea-salt $SO_4^=$, they use concentrations of 0.2 μg to 0.4 μg S/m^3 and a deposition velocity of 1.5 ± 0.5 cm/sec.

Savoie (1984), using the data base discussed in the previous section on $SO_4^=$ wet-deposition and a dry-deposition velocity of 0.05 cm/sec (corrected for the distribution of excess $SO_4^=$ as a function of particle radius), estimates a dry-deposition rate for excess $SO_4^=$ of 0.011 g $S/(m^2 \cdot yr)$.

Mészáros (1982) estimates values of 0.12 g and 0.02 g $S/(m^2 \cdot yr)$ for rates of the dry deposition of sea-salt $SO_4^=$ and excess $SO_4^=$, respectively. The deposition of both types of $SO_4^=$ is controlled by larger particles. The estimate for sea-salt $SO_4^=$ is based on a concentration of 0.3 μg S/m^3. For excess $SO_4^=$, Mészáros (1982) uses a total concentration of 0.3 μg S/m^3, which is, however, distributed between two particle-size classes. On larger sea-salt particles ($r > 0.5$ μm), the concentration of excess $SO_4^=$ is 0.19 μg S/m^3 and, on the smaller ones ($r < 0.5$ μm), the concentration is 0.11 μg S/m^3. The deposition velocities Mészáros uses are not given but are selected as a function of particle size and wind speed from Sehmel and Sutter (1974).

Both the concentrations and the deposition velocities used to calculate dry deposition for sea-salt $SO_4^=$ and excess $SO_4^=$ vary greatly (Table 8-2), with estimates of the dry deposition of sea-salt $SO_4^=$ ranging from 0.02-0.2 g $S/(m^2 \cdot yr)$ and, for excess $SO_4^=$, from 0.006-0.033 g $S/(m^2 \cdot yr)$. For purposes of this discussion, I selected approximate midrange values of 0.10 g and 0.02 g $S/(m^2 \cdot yr)$ for the dry-deposition rates for sea-salt $SO_4^=$ and excess $SO_4^=$, respectively, with the realization that the uncertainties around this value are probably near a factor of ten (Fig. 8-1).

8.2.4. Continental Areas: Wet Deposition

8.2.4.1. **Sulfate**. Ryaboshapko (1983) has published the most recent compilation of estimates of S deposition to remote continental areas. For all continents he assumes an average S concentration of 44 μeq/l and a deposition of 0.28 ± 0.08 g $S/(m^2 \cdot yr)$, using an annual precipitation rate of 40 cm. This estimate can be compared to another estimate of S in wet deposition to remote continental areas performed by Granat et al. (1976). In that paper the estimate, by difference, is 0.15 g $S/(m^2 \cdot yr)$, lower than the 0.28 g $S/(m^2 \cdot yr)$ estimated by Ryaboshapko (1983). To assess which of the estimates is more realistic, I examined recent data from the GPCP and other programs.

The compilation of such studies (Table 8-3) shows that the range of the $SO_4^=$ concentrations (2.7-13 μeq/l) is substantially lower than the 44 μeq/l estimated by Ryaboshapko (1983). I felt his estimate was too high, probably because of sampling and analytical problems. I have already discussed the difficulties of collecting samples on other than an event basis. In addition to these problems there are special problems associated with continental samples. The most important of these is the strict limitation placed on the accuracy of the analyses because of problems of analytical sensitivity and contamination.

The three common analytical techniques used to measure $SO_4^=$ and the errors associated with them are (1) gravimetric with an error of $\pm$ 20 μeq/l, (2) colorimetric with an error of $\pm$ 4 μeq/l, and (3) ion chromatographic with an error of $\pm$ 1 μeq/l. Given the low $SO_4^=$ concentrations in Table 8-3, ion chromatography is the preferred technique for measuring $SO_4^=$ in wet deposition in remote continental areas because of its greater sensitivity and fewer interferences.

Although Ryaboshapko's (1983) estimate of the $SO_4^=$ concentration in wet deposition seems to be rather high, his estimate of the wet deposition rate of $SO_4^=$ to remote continental areas (0.28 g $\pm$ 0.08 g S/[m^2 · yr]) is within the range of other studies (Table 8-3). This could be because Ryaboshapko's estimate for the precipitation rate (40 cm/yr) may be too low. A more reasonable continental value would be 50-100 cm/yr (Fig. 8-3). His value of 40 cm/yr for the world's continents is probably only representative of the higher latitudes. In temperate and tropical regions, the precipitation amounts are nearer to 200 cm/yr.

The value of 0.4 g S/(m^2 · yr) from Stallard and Edmond (1981) for the Amazon is probably an upper limit because their sampling period is confined to May and June, the dry season. Therefore, their estimates for $SO_4^=$ concentrations could be higher than the annual weighted average. This idea is supported by data from San Carlos, Venezuela (Galloway et al. 1982), and Lake Calado, Brazil (unpublished data available from Dr. John Melack, University of Santa Barbara, California), which cover an entire year (Table 8-3). The estimates from these data of $SO_4^=$ in wet

Table 8-3. Wet Deposition of $SO_4^=$ to Remote Continental Areas.

Location	No. of Samples	Precipitation Amount (cm)	$SO_4^=$ Concentration (μeq/l)	$SO_4^=$ Deposition (g S/[m^2·yr])	Reference
Global Estimates	–	40	44	0.28±0.08	Ryaboshapko 1983
	–	–		0.15	Granat et al. 1976
East Africa	61	–	13	0.15	Rodhe et al. 1981
Nigeria	–	–	8	0.11	Bromfield 1974
Venezuela					
San Carlos	14	391	2.7	0.17	Galloway et al. 1982
Brazil					
Amazon Basin	28	240	8.8	0.4	Stallard & Edmond 1981
Lake Calado	59	200	3.8	0.12	Melack unpubl. data
Australia					
Katherine	120	112	4.7	0.064	Likens et al. 1985

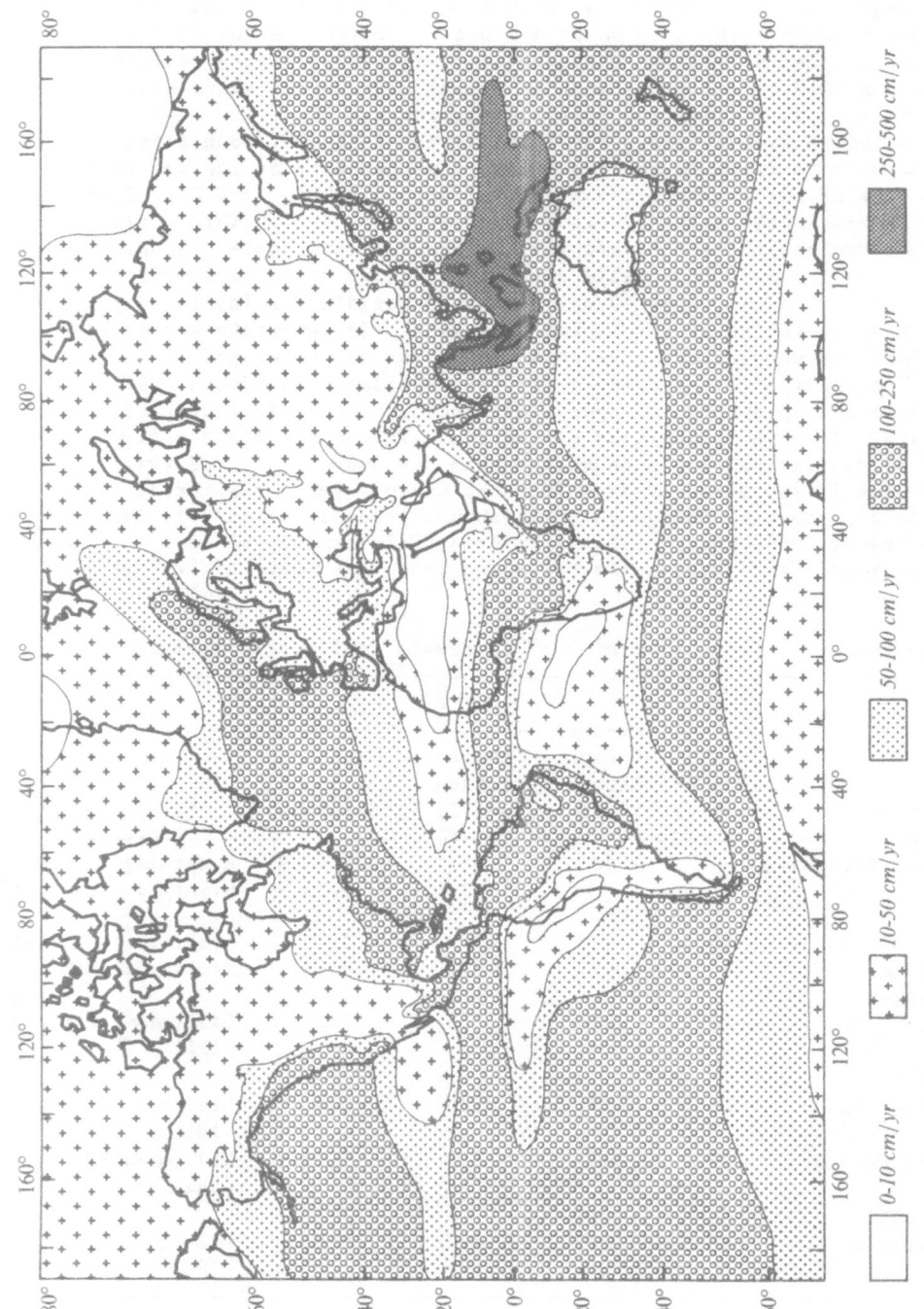

Figure 8–3. Global Rates of Wet Deposition (cm) (Lamb 1972).

0-10 cm/yr

10-50 cm/yr

50-100 cm/yr

100-250 cm/yr

250-500 cm/yr

deposition (0.17 g and 0.12 g S/[m^2 · yr], respectively) are significantly lower than the estimates of Stallard and Edmond (1981). Investi-
gations of the SO$_4^=$ concentration in precipitation from Africa yield two
estimates: 0.15 g and 0.11 g S/(m^2 · yr) from East Africa and Nigeria,
respectively (Rodhe et al. 1981, Bromfield 1974). These values agree
well with those from South America. Using 120 samples collected in
Katherine, Australia, Likens and his colleagues (1985) estimate a wet-
deposition rate for SO$_4^=$ of 0.064 g S/(m^2 · yr) (Fig. 8-1). This value
is the lowest of the range and probably near the lower limit for the
annual rate of SO$_4^=$ deposition to remote areas.

Since the estimates of SO$_4^=$ in wet deposition to remote continental
areas range from 0.064–0.4 g S/(m^2 · yr) (Table 8-3), I selected, for
comparison purposes, a value of 0.2 g S/(m^2 · yr). Given the paucity of
data, this value should be considered the median of a large range.

8.2.5. Continental Areas: Dry Deposition

8.2.5.1. Sulfur Dioxide. Estimates of the dry-deposition rates of SO$_2$
to remote continental areas range from 0.02 g to 0.4 g S/(m^2 · yr)
(Table 8-4). Möller's (1983) estimate of 0.1 g S/(m^2 · yr) uses an
atmospheric concentration of 0.4 μg S/m^3 and a deposition velocity of
0.8 cm/sec. Ryaboshapko (1983) has estimated a value of 0.04 g ± 0.02 g
S/(m^2 · yr) from atmospheric concentrations of 0.2 ± 0.1 μg S/m^3 and a
deposition velocity of 0.6 ± 0.2 cm/sec. Granat (1976), using a different method by assuming that dry deposition is 25% of wet deposition,

Table 8-4. Concentrations, Deposition Velocities, and Dry-deposition
Rates for Sulfur Species in Remote Continental Areas.

Atmospheric Concentration (μg S/m^3)	Deposition Velocity (cm/sec)	Dry-Deposition Rate (g S/[m^2 · yr])	Reference
		SO$_4^=$	
0.6±0.2	0.2	0.04 (0.02–0.13)	Ryaboshapko 1983
0.5	0.1	0.02	Möller 1983
1±0.5	0.1±0.05	0.01–0.07	Janssen-Schmidt et al. 1981
		SO$_2$	
0.2±0.1	0.6±0.2	0.04±0.02	Ryaboshapko 1983
*	*	0.03	Granat 1976
0.4	0.8	0.1	Möller 1983
0.5–1.0	0.8±0.4	0.06–0.4	Janssen-Schmidt et al. 1981

*Granat (1976) calculated dry deposition by assuming it is 25% of
wet deposition.

arrives at an estimate of 0.03 g S/(m^2 · yr). From the few continental
data from the Soviet Union, Africa, Australia, and Canada that are compiled by Janssen-Schmidt et al. (1981), a flux of 0.06-0.4 g S/(m^2 · yr)
is calculated (Table 8-4).

Given the small size of the data base on SO$_2$ concentrations in
remote areas and the uncertainties of the deposition velocities, the
agreement between these values should not be taken too seriously. For
purposes of this discussion, I chose a dry-deposition rate of SO$_2$ to
remote continental areas of 0.07 g S/(m^2 · yr). However, because of
differences in vegetation type, surface wetness, temperature, and other
factors, the variability about this single number was certainly large.

8.2.5.2. <u>Sulfate</u>. Ryaboshapko (1983), assuming atmospheric concentrations of 0.6 ± 0.2 µg S/m^3 and a deposition velocity of 0.2 cm/sec
(range = 0.1-0.5 cm/sec), calculates a dry-deposition rate of 0.04 g
S/(m^2 · yr) (range 0.02-0.13 g S/[m^2 · yr]) for SO$_4^=$ to remote continental areas. Möller (1983) calculates 0.02 g S/(m^2 · yr) deposited using
an atmospheric SO$_4^=$ concentration of 0.5 µg S/m^3 and a deposition velocity of 0.1 cm/sec (Table 8-4). Based on data from remote continental
regions, Janssen-Schmidt et al. (1981) estimate a dry-deposition rate of
0.01-0.07 g S/(m^2 · yr).

These three estimates agree closely but, again like the others, they
would be improved by additional data. For purposes of this discussion,
the value of 0.03 ± 0.02 g S/(m^2 · yr) for SO$_4^=$ in dry deposition in
remote continental areas appears to be reasonable.

8.2.6. Polar Areas

Polar areas are defined as those areas north and south of 65° N and 65° S
and in this study are represented by Greenland and Antarctica, respectively. The deposition of S to these areas represents a special case.
Not only are the precipitation rates substantially lower than for most
other remote areas but also the lack of sources for S species in the
atmosphere ensures that both the concentrations of S species in atmospheric deposition and the deposition rates from the atmosphere will be
much lower than for other remote areas. The following sections summarize
our limited knowledge of the atmospheric deposition of S to these areas.
With the exception of one study for the dry deposition of SO$_2$, wet and
dry deposition are discussed together since the samples collected from
ice cores or the snow surface represent total deposition.

8.2.6.1. <u>Excess Sulfate</u>. In the studies of the deposition of excess
SO$_4^=$ to polar areas (Table 8-5), concentrations range from 0.5 µeq/l to
2.5 µeq/l. For these 21 studies, the mean and standard deviation of the
excess SO$_4^=$ concentrations is 1.2 ± 0.5 µeq/l.

There are no significant differences between the means from the
southern and the northern polar areas with regard to deposition rates.
The means and standard deviations are 0.0029 ± 0.0033 g S/(m^2 · yr) and
0.0047 ± 0.0041 g S/(m^2 · yr) for the southern and northern polar areas,
respectively. Although the mean of the southern data is less than that
of the northern, the variances about both means are large enough to

Table 8-5. Wet Deposition of Excess Sulfate to Polar Areas.

| Location | Precipitation Accumulation Rate (cm/yr) | Excess $SO_4^=$ | | Reference |
		Concentration (μeq/1)	Deposition (g S/[m^2 · yr])	
Greenland		0.98±0.13	0.0019±0.0003	Busenberg & Langway 1979
North Central	13	1.55	0.0032	Herron 1982
Crete	27	0.7	0.0030	Herron 1982
Camp Century	32	0.75	0.0038	Herron 1982
Camp Century	32	1.7	0.0033	Cragin et al. 1974
Dye 3	50	0.5	0.0040	Herron 1982
Milcent	50	1.7	0.014	Herron et al. 1977
Antarctica	12	1.6	0.0031	Delmas 1982
East	–	1.5	0.0029	Delmas & Boutron 1978
Central	–	1.2	0.0023	Delmas & Boutron 1980
Vostok	2.2	2.5	0.0009	Doronin 1975
Dome C	3.7	1.4	0.0008	Legrand & Delmas 1985
South Pole	8.0	0.9	0.0011	Legrand & Delmas 1984
D 55	8.3	1.0	0.0013	Legrand & Delmas 1985
Ross Ice Shelf	10	1.2	0.0019	Herron 1982
Byrd Station	16	1.2	0.0031	Cragin et al. 1974
Byrd Station	16	0.7	0.0018	Herron 1982
D 80	23	0.6	0.0022	Legrand & Delmas 1985
D 57	27	0.42	0.0018	Legrand & Delmas 1985
Siple	40	0.6	0.0038	Herron 1982
J. Ross Island	60	1.5	0.014	Aristarain et al. 1982

eliminate any significant difference. The high degree of variation is primarily caused by large spatial differences in precipitation amount as the concentrations in deposition are essentially identical.

For purposes of comparison to other fluxes, I chose a value for the wet deposition of excess $SO_4^=$ to polar areas of 0.003 g S/(m^2 · yr).

8.2.6.2. _Sea-salt_ _Sulfate_. Deposition rates of sea-salt $SO_4^=$ are generally lower than those of excess $SO_4^=$ except in coastal areas where sea-salt deposition rates increase (Ryaboshapko 1983). In midpolar regions, the deposition of sea-salt $SO_4^=$ represents only 10% of the total sulfate (e.g., Antarctic, Delmas et al. 1982b).

8.2.6.3. _Dry_ _Deposition_: _Sulfate_. The dry-deposition rates of excess and sea-salt $SO_4^=$ are largely unmeasured in polar areas. Again, near the coast, the dry-deposition rates of sea-salt $SO_4^=$ are expected to become more important than the dry-deposition rates of excess $SO_4^=$.

8.2.6.4. Dry Deposition: Sulfur Dioxide.

Delmas (1982) estimates that the dry-deposition rate of SO_2 is 0.001 g S/(m^2 · yr) in the Antarctic, based on an SO_2 concentration of 0.05 µg S/m^3 and a deposition velocity to snow of 0.08 cm/sec.

8.2.7. Summary

The estimates for the wet and dry deposition of S species to remote marine and continental areas are summarized in Figure 8-1. In combination with the data from polar areas, there are several points to be made.

1. In remote marine, continental, and polar areas, the wet deposition of excess S is usually more than twice that of the dry deposition. This partition, however, depends on the rainfall rate in each region. In some remote continental areas, such as deserts, the dry deposition may be the most important removal mechanism.

2. In remote marine areas, the wet deposition of sea-salt $SO_4^=$ appears to be several times greater than that of the dry deposition and the dry deposition of SO_2 and excess $SO_4^=$ are of similar magnitude

4. In remote continental areas, the dry deposition of SO_2 can be twice that of $SO_4^=$.

5. In polar areas, the wet deposition of excess $SO_4^=$ is greater than the dry deposition.

8.3. NITROGEN

8.3.1. Introduction

This discussion of the atmospheric deposition of nitrogen to remote areas considers gaseous NO, NO_2, HNO_3, NH_3, and aerosol NO_3^- and NH_4^+. As with S, three types of programs have been used to determine N in wet deposition to remote areas—long-term programs that sample atmospheric wet deposition on a monthly basis (e.g., WMO), short-term programs that sample atmospheric deposition on an event basis (e.g., SEAREX), and long-term programs that sample atmospheric wet deposition on an event basis (GPCP). In addressing atmospheric wet deposition to remote areas, data evaluation from long-term programs that only sample on a monthly basis is a difficult task. Not only do the data on the N species suffer from problems similar to those that affect the S species (i.e., evaporation and contamination) but there are also potential problems with the biological uptake and transformation of the nitrogen compounds. Therefore, this analysis emphasizes the event sampling of atmospheric wet deposition during short- and long-term programs.

Estimates of the rates of dry deposition are of two types—those from calculations based on atmospheric concentrations and deposition velocities and those from measurements using surrogate surfaces.

8.3.2. Marine Areas: Wet Deposition

8.3.2.1. Nitrate. A recent compilation by Logan (1983b) estimates that the NO_3^- deposition in remote marine areas is 0.01-0.03 g N/(m^2 · yr). This range of values agrees well with an earlier one by Söderlund and Svensson (1976) of 0.01-0.04 g N/(m^2 · yr). Logan's estimate is based on short-term sampling programs at Samoa (0.005 g N/[m^2 · yr], Pszenny et al. 1982) and Hawaii (0.015 g N/[m^2 · yr], unpublished data available from Dr. John Miller, Air Resources Laboratory, NOAA, Silver Spring, Maryland) and on an 11-month data set covering 26 events from Amsterdam Island in the Indian Ocean (0.028 g N/[m^2 · yr], Galloway and Gaudry 1984). Since the data base used in Logan's analyses is small, it is

Table 8-6. Wet Deposition of NO_3^- and NH_4^+ (g N/[m^2 · yr]) to Remote Marine Areas.

Location	No. of Samples	NO_3^-	NH_4^+	References
Global Estimates	–	0.01-0.03	0.01-0.03	Logan 1983b, Galloway et al. 1982
	–	0.01-0.04	0.02-0.07	Söderlund & Svensson 1976
	–	0.020		Savoie 1984
		0.026±0.02	0.02±0.02	Böttger et al. 1978
Pacific Ocean				
Samoa	12	0.005	–	Pszenny et al. 1982
Hawaii		0.005-0.032	–	Miller, unpbl. data
New Zealand	7	0.008		Pszenny & Duce, unpbl. data
RV Korolev	6	0.03	0.062	Galloway, unpbl. data
RV Discoverer	4	0.05	0.060	Galloway, unpbl. data
Central Tasman Sea	4	0.04		Pszenny & Duce, unpbl. data
Indian Ocean				
Amsterdam Island	66	0.021	0.028	Galloway & Gaudry 1984
Western Atlantic Ocean*				
Bermuda**	19	0.10	0.050	Galloway, unpbl. data
Shipboard**	27	0.07	0.062	Galloway et al. 1983, Galloway, unpbl. data

*Deposition was calculated using a precipitation rate of 150 cm/yr.

**Only storms associated with air masses originating east of 60° W.

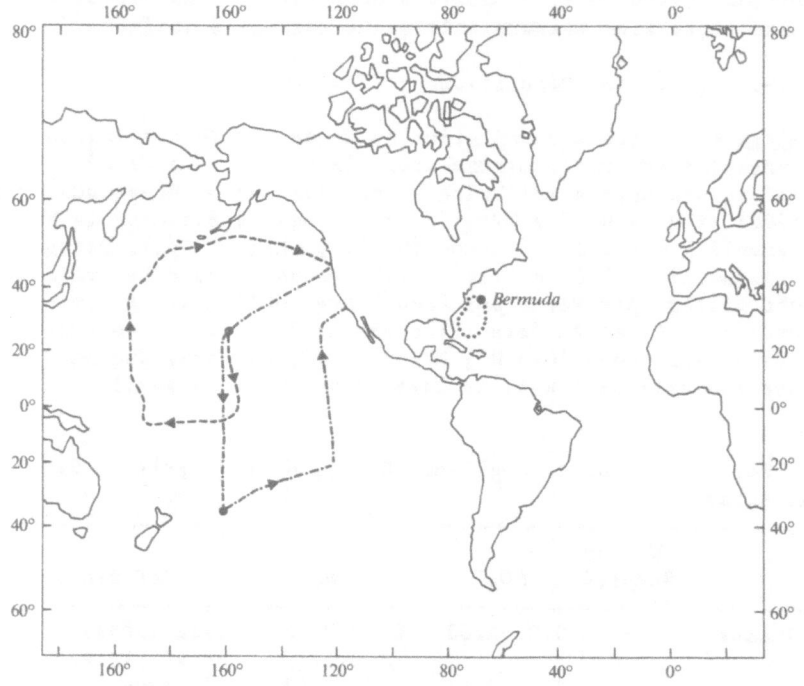

Figure 8-4.   Cruise tracks of the RV Akademic Korolev (-·-·-), the
    RV Discoverer (---), and the Homes Lines ships (····), the ships
    used to collect wet deposition.

instructive to compare these data (Table 8-6) to more recent data sets
that have been collected.
      Six rain samples have been collected aboard the RV Korolev and four
aboard the RV Discoverer when they were in the Pacific Ocean (unpublished
data available from the author, Fig. 8-4). The average NO_3^- concentrations are 1.3 µeq/l from the Korolev samples and 2.2 µeq/l from the
Discoverer samples. The average NO_3^- depositions are 0.03 g and 0.05 g
$N/(m^2 \cdot yr)$, respectively, assuming an annual precipitation of 150 cm.
In the Atlantic Ocean, 19 storms associated with air masses originating
east of 60° W have been sampled on Bermuda over a 2-year period (unpublished data available from the author). These samples have a mean NO_3^-
concentration of 4.5 µeq/l. Assuming that the annual precipitation is
150 cm/yr, this equates to a deposition of 0.10 g $N/(m^2 \cdot yr)$. Similarly, samples from 27 storms associated with easterly air masses have
been collected onboard three cruise ships in the Western Atlantic Ocean
between May-November of 1981, 1982, and 1983 (Galloway et al. 1983 and
unpublished data available from the author, Fig. 8-4). The average NO_3^-

concentration of these storms is 3.1 µeq/l, which equates to an annual
deposition of 0.07 g N/(m^2 · yr) using an annual precipitation of
150 cm/yr.

Pszenny and Duce (unpublished data available from A. A. P. Pszenny,
Department of Environmental Sciences, University of Virginia, Charlottesville) have found that the concentrations of NO_3^- in wet deposition at
New Zealand (90–Mile Beach) and in the Central Tasman Sea are 0.5 µeq/l
and 2.2 µeq/l, respectively. The NO_3^- deposition values are 0.008 g and
0.04 g N/(m^2 · yr), respectively. Using an extensive data base of
aerosol concentrations, Savoie (1984) estimates that the average wet–
deposition rate of NO_3^- to the global oceans is 0.020 g N/(m^2 · yr).
Böttger et al. (1978) use an extensive survey of NO_3^- wet deposition over
marine areas as the basis for their estimate of 0.026 ± 0.02 g N/(m^2 ·
yr) (Table 8–6).

Logan (1983b) cites a mean annual wet–deposition rate for NO_3^- of
0.027 g N/(m^2 · yr) on Amsterdam Island based on the 26 storms reported
in Galloway et al. (1982). Galloway and Gaudry (1984) have reported an
updated estimate based on an expanded data set (66 storms) of 0.021 g
N/(m^2 · yr), which agrees fairly well with the earlier estimate.

The range of estimates of NO_3^- deposition is 0.005–0.10 g N/(m^2 ·
yr). Most of the studies agree with Logan's (1983b) estimate of 0.01 g
to 0.03 g N/(m^2 · yr). There are several possible reasons for the wide
range. Several of the estimates are based on small data bases that may
be seasonally biased. Some of the higher numbers may be influenced by
anthropogenic activities (e.g., the Western Atlantic data). In addition,
since the continents are stronger sources for NO_x than are oceans (see
Chapter 3), there would be a spatial variability of NO_3^- in wet deposition that is controlled by distance from continents.

A value of 0.03 g N/(m^2 · yr) has been selected for comparison with
other fluxes with the realization that there may have been a large amount
of variability about this value (Fig. 8–5).

8.3.2.2. **Ammonium.** Söderlund and Svensson (1976) and Logan (1983b)
estimate that wet deposition of NH_4^+ to remote marine areas ranges from
0.02–0.07 g and 0.01–0.03 g N/(m^2 · yr), respectively. Böttger et al.
(1978) estimate a rate of 0.02 ± 0.02 g N/(m^2 · yr). As I did with NO_3^-,
I have used even more recently gathered data to evaluate these estimates
(Table 8–6). The sources of the more recent data are the same as for
NO_3^-: two ships in the Pacific Ocean, Bermuda, and three cruise ships in
the Western Atlantic (tracks in Fig. 8–4). In addition, I have used a
data set on wet deposition from Amsterdam Island in the Indian Ocean.

In the Pacific Ocean, the averages of the NH_4^+ deposition onboard
each of the ships are 0.060 g and 0.062 g N/(m^2 · yr) (unpublished data
available from the author). These estimates are several times higher
than Logan's (1983b) and Böttger et al. (1978) but agree well with the
upper limit of Söderlund and Svensson (1976) (Table 8–6).

The new data for the Atlantic Ocean (considering only storms that
originated east of 60° W) show an NH_4^+ flux of 0.050 g N/(m^2 · yr) for 19
events collected on Bermuda and 0.062 g N/(m^2 · yr) for 27 events collected onboard three cruise ships (Galloway et al. 1983 and unpublished
data available from the author). Again these data are higher than

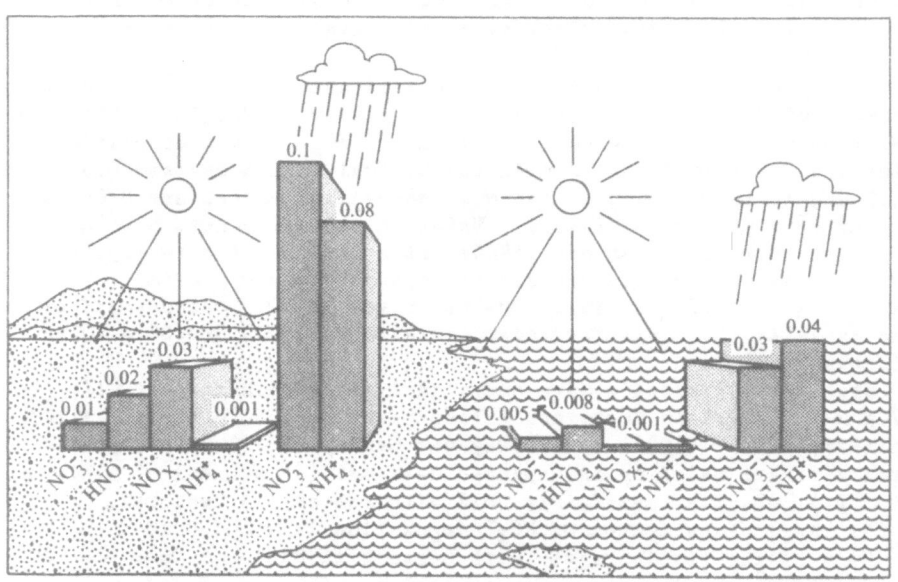

Figure 8-5. Wet- and Dry-deposition Rates of N Species to Remote Areas (g N/[m² · yr]).

Logan's (1983b) and Böttger et al.'s (1978) but fall within the upper
limits of Söderlund and Svensson's (1976) estimate. However, in contrast
to NO_3^-, these values are not significantly higher than those from the
new data for the Pacific Ocean. This indicates that there does not appear
to be an extra source of NH_4^+ in the Atlantic Ocean atmosphere, as is the
case with NO_3^- (Table 8-6).

For the Indian Ocean, based on a more extensive data set (N=66)
from Amsterdam Island (Galloway and Gaudry 1984), the wet-deposition rate
of NH_4^+ is 0.028 g N/(m² · yr), which is not significantly different from
the 0.033 g N/(m² · yr) used in Logan (1983b) and based on a smaller data
set (N=26) (Table 8-6).

The results of this update on the wet deposition of NH_4^+ to remote
marine areas are contradictory. The new Indian Ocean data (0.028 g N/[m²
· yr]) from Galloway and Gaudry (1984) agree with the earlier estimates
of Logan (1983b) of 0.033 g N/(m² · yr) (based on Galloway et al. 1982)
and Bött-ger et al. (1978) but are less than those of Söderlund and
Svensson (1976). Estimates based on the new data from both the Pacific
Ocean (0.060-0.062 g N/[m² · yr]) and Atlantic Ocean (0.050-0.062 g
N/[m² · yr]) agree well with the upper limit of Söderlund and Svensson's
global estimates (0.02-0.07 g N/[m² · yr]) but are higher than the upper
limit of the estimate of Logan (1983b) and Böttger et al. (1978) by
about a factor of two. To resolve this discrepancy, additional data from
remote marine areas are required. For purposes of comparison with other
fluxes, I selected a value of 0.04 g N/(m² · yr) (Fig. 8-5).

8.3.3. Marine Areas: Dry Deposition

8.3.3.1. Nitrate. Logan (1983b) gives an estimate of 0.006 g N/(m^2 · yr) for the dry-deposition rate of NO_3^- aerosol to remote marine areas. Söderlund and Svensson (1976) estimates this rate to be 0.0003 g N/(m^2 · yr). The former estimate uses higher values of both NO_3^- concentrations and deposition velocities (Table 8-7). The higher value of the atmospheric concentration is apparently justified since it is based on several recent reports (as cited in Logan 1983b). The higher value for the

Table 8-7. Concentrations, Deposition Velocities, and Dry-deposition
Rates for Nitrogen Species in Remote Marine Areas.

Atmospheric Concentration (μg N/m^3)	Deposition Velocity (cm/sec)	Dry- Deposition Rate (g N/[m^2 · yr])	Reference
		NO$_3^-$	
0.06	0.3	0.006	Logan 1983b
0.02	0.05-0.15	0.0003	Söderlund & Svensson 1976
0.067*	0.25	0.005*	Savoie 1984
-	-	0.01	Pszenny & Duce unpbl. data
		HNO$_3$	
0.03	0.6-1.0	0.006-0.01	Logan 1983b
0.01	0.5	0.002	Böttger et al. 1978
		NO$_x$	
0.2	0.3-0.8	0.02-0.05	Söderlund & Svensson 1976
0.06	0.3-0.6	0.006-0.01	Logan 1983b
0.03	0.015	0.00014	Böttger et al. 1978
		NH$_3$	
0.1	0.001	0.003	Georgii & Gravenhorst 1977
0.06	0.8	0.02	Söderlund & Svensson 1976, Ayers & Gras 1983
0.0006-0.06**	0.8	0.0002-0.02	Taylor et al. 1983
0.06-0.6+	0.8	0.02-0.2	Taylor et al. 1983
		NH$_4^+$	
0.4	0.03	0.003-0.01	Söderlund & Svensson 1976
<0.01-0.2	-	-	Huebert & Lazrus 1980

*World ocean's average.

**Range of NH$_3$ concentration in equilibrium with acidic, wet SO$_4$ aerosols.

+Range of NH$_3$ concentration in equilibrium with ocean surface.

deposition velocity is based on recent determinations of the variation of NO_3^- concentration as a function of particle size. Specifically, a large portion of the NO_3^- appears to be associated with a larger diameter sea-salt aerosol, hence the higher deposition velocity is justified (Gravenhorst 1975, Gravenhorst et al. 1979, Savoie and Prospero 1982). Pszenny and Duce (unpublished data available from A. A. P. Pszenny, Department of Environmental Sciences, University of Virginia, Charlottesville) estimate a NO_3^- dry deposition of 0.01 g N/(m^2 · yr) based on seven samples collected on a teflon plate at 90-Mile Beach, New Zealand. Given the differences in methodology, this factor-of-two agreement is remarkable. As additional support for the higher values, Savoie (1984), after analyzing over 1,000 bulk aerosol samples, has concluded that the global average atmospheric NO_3^- concentration is 0.067 µg N/m^3. When coupled with deposition velocity of 0.25 cm/sec, this gives a NO_3^- dry-deposition rate of 0.005 g N/(m^2 · yr).

For the purposes of calculation, I have selected a value of 0.005 g N/(m^2 · yr) for the dry-deposition rate of NO_3^- aerosol to remote marine areas (Table 8-7).

8.3.3.2. Nitric Acid.
Logan (1983b) estimates a rate of dry deposition for HNO_3 in remote areas of 0.006-0.01 g N/(m^2 · yr) (Table 8-7). Böttger et al. (1978) estimate a value of 0.002 g N/(m^2 · yr). Given the problems of sampling and analyzing low concentrations of HNO_3, it is impossible to judge these data. Additional data are certainly required.

For purposes of comparisons with other fluxes, a value of 0.008 g N/(m^2 · yr) was selected for the dry-deposition rate of HNO_3 to remote marine areas.

8.3.3.3. Nitrogen Oxides.
NO_x is defined as $NO + NO_2$. However, measurements to determine concentrations of NO_x can include PAN and HNO_3 besides NO and NO_2 (Spicer 1982, Winer et al. 1974). Logan (1983b) has summarized recent measurements of NO_x in remote areas. From that summary, she suggests an average NO_x concentration of 0.06 µg N/m^3. Using this value and a deposition velocity ranging from 0.3-0.6 produces a dry-deposition rate for NO_x of 0.006-0.01 g N/(m^2 · yr). This can be compared to the dry-deposition rate of 0.02-0.05 g N/(m^2 · yr) estimated by Soderlund and Svensson (1976) (Table 8-7). The latter is higher because of the selection of a much larger NO_x concentration in remote marine atmospheres. Böttger et al. (1978) estimate a NO_2 dry-deposition rate of 0.00014 g N/(m^2 · yr) using an atmospheric concentration of 0.03 µg N/m^3 and a deposition velocity of 0.015 cm/sec. They have also stated that NO deposition can be ignored because NO is adsorbed by the ocean surface much slower (by about a factor of 10) than NO_2.

Thus NO_x dry deposition ranges from 0.00014 g to 0.05 g N/(m^2 · yr). Clearly additional data are required on both the concentrations and the deposition velocities of NO and NO_2 in the marine-boundary layer. For the sake of comparison with other fluxes, I selected a value of 0.001 g N/(m^2 · yr) (Fig. 8-5).

8.3.3.4. Ammonia.
The concentration of NH_3 in the remote marine atmosphere seems to be quite variable (Table 8-7). If the ammonia gas-phase

concentration is being controlled by equilibrium with the ocean surface, an NH_3 concentration in the marine-boundary layer below about 1 µg N/m^3 would be suggested, according to temperature, pH value, and ammonium concentration of the surface water (Georgii and Gravenhorst 1977, Ayers and Gras 1983). Deviations from equilibrium conditions would result in NH_3 fluxes into or out of the ocean water. The direction and magnitude of NH_3 fluxes across the ocean/atmosphere interface in different regions of oceans are not yet known.

8.3.3.5. Ammonium.

Söderlund and Svensson (1976) have estimated a dry-deposition rate of NH_4^+ aerosol of 0.003–0.01 g $N/(m^2 \cdot yr)$, which is based on a deposition velocity of 0.03 cm/sec and an atmospheric concentration of 0.4 µg N/m^3 (Table 8-7). However, Huebert and Lazrus (1980) state that the concentration range of NH_4^+ aerosol in the marine-boundary layer is < 0.01–0.2 µg N/m^3. In addition, since marine NH_4^+ is concentrated in the accumulation mode (r < 0.5 µm), the dry-deposition velocity should be in the range of 0.01–0.1 cm/sec (Gravenhorst 1975, 1978). Therefore using a range of atmospheric NH_4^+ concentration of 0.1–0.4 µg N/m^3 (Table 8-7) and a deposition-velocity range of 0.01 cm/sec to 0.1 cm/sec, a dry-deposition rate ranging from 0.0003 g to 0.01 g $N/(m^2 \cdot yr)$ is calculated. For the purpose of comparison with other fluxes, I have selected a dry-deposition rate of NH_4^+ from the remote marine atmosphere of 0.001 g $N/(m^2 \cdot yr)$ (Fig. 8-5).

8.3.4. Continental Areas: Wet Deposition

Logan's 1983b article contains the most recent estimates of wet deposition of NO_3^- and NH_4^+ to remote continental areas. These estimates (0.05 g $N/[m^2 \cdot yr]$) are primarily based on the data from three continental sites in the Global Precipitation Chemistry Project in San Carlos, Venezuela; Katherine, Australia; and Poker Flat, Alaska, reported by Galloway et al. (1982). In the next two sections, I will compare these estimates to additional data from Katherine, Australia, and to data from other studies.

8.3.4.1. Nitrate.

The data base from Australia reported by Galloway et al. (1982) and used by Logan (1983b) has recently been considerably expanded (Likens et al. 1985). Using the revised data base, the wet deposition rate of NO_3^- in Australia is calculated to be 0.077 g $N/(m^2 \cdot yr)$ (Table 8-8). The new Australian estimates are similar to those used by Logan, 0.055 g $N/(m^2 \cdot yr)$, indicating that 0.07 g $N/(m^2 \cdot yr)$ is probably a good estimate for that site.

    Other data bases that provide data on the wet deposition of NO_3^- to remote continental areas are Stallard and Edmond (1981) and Melack (unpublished data available from Dr. J. Melack, University of Santa Barbara, California). Stallard and Edmond (1981) have sampled 28 precipitation events in the Amazon basin of Brazil and record a mean concentration of 2.1 µeq/l. Using an annual precipitation rate of 391 cm (the same as used at San Carlos, Venezuela), an annual deposition rate of 0.11 g $N/(m^2 \cdot yr)$ is calculated (Table 8-8). Melack has calculated a wet-deposition rate of NO_3^- of 0.10 g $N/M^2/y$. Both estimates agree well

Table 8-8. Wet Deposition of NO_3^- and NH_4^+ to Remote Continental Areas.

Location	Number of Samples	Annual Precipitation (cm)	NO_3^- Precipitation Concentration (µeq/l)	Deposition (g N/[m²·yr])	NH_4^+ Precipitation Concentration (µeq/l)	Deposition (g N/[m²·yr])	Reference
Global Estimates							
	–	–	–	0.10-0.23	–	0.23-0.47	Söderlund & Svensson 1976
	–	–	–	0.05*	–	0.05*	Logan 1983b
	–	–	–	0.015-0.05	–	0.01-0.06	Böttger et al. 1978
Venezuela							
San Carlos	14	391	2.6	0.14	2.3	0.13	Galloway et al. 1982, Logan 1983b
Brazil							
Amazon basin	28	391**	2.1	0.11	1.1	0.062	Stallard & Edmond 1981
Lake Calado	54	200	3.5	0.10	3.1	0.087	Melack, unpubl. data
Australia							
Katherine	40	112	4.3	0.055	2.0	0.033	Galloway et al. 1982, Logan 1983b
	120	112	4.9	0.077	3.4	0.056	Likens et al. 1985

*Lower estimate.

**Estimated.

with the 0.14 g N/(m^2 · yr) value from the GPCP site in Venezuela. They are all about a factor of two higher than the world-wide distribution of 0.015-0.05 g N/(m^2 · yr) estimated by Böttger et al. (1978) and near the range of 0.10-0.23 g N/(m^2 · yr) reported by Söderlund and Svensson (1976).

Thus, the available data show that the deposition of NO_3^- to remote continental areas covers a range of 0.015 g to 0.23 g N/(m^2 · yr). Without additional data from remote continental forest regions, it is impossible to determine whether Logan's (1983b) estimate of 0.05 g N/(m^2 · yr) or Böttger et al.'s (1978) estimate of 0.015-0.05 g N/(m^2 · yr) is the more reasonable value to use as a global average or if the value should be higher. For purposes of comparison with other fluxes, I chose a value of 0.1 g N/(m^2 · yr) (Fig. 8-5).

8.3.4.2. **Ammonium**. Logan (1983b) estimates a global value of 0.05 g N/(m^2 · yr) for NH_4^+ in wet deposition to remote continental areas. Böttger et al. (1978) estimate a global range of 0.01-0.06 g N/(m^2 · yr). Söderlund and Svensson (1976) report a range of NH_4^+ deposition of 0.23-0.47 g N/(m^2 · yr). Logan's (1983b) estimate is based on preliminary data from the Global Precipitation Chemistry Project (Galloway et al. 1982). Specifically, for sites in Australia and Venezuela, the annual wet-deposition rates of NH_4^+ are reported as 0.033 g and 0.13 g N/(m^2 · yr), respectively. The update of the data from the Australian site (Likens et al. 1985) yields an annual wet-deposition rate of 0.056 g N/(m^2 · yr) (Table 8-8). The new value from Likens et al. is about a factor of two greater than the older values from Katherine, Australia.

Additional data presented in Stallard and Edmond (1981) and by Melack (unpublished data available from Dr. J. Melack, University of Santa Barbara, California) for the Amazon rain forest in Brazil and for Lake Calado, Brazil, indicate that the annual deposition of NH_4^+ is 0.062 g and 0.087 g N/(m^2 · yr), respectively, or about half of what is reported using the smaller data base collected at San Carlos, Venezuela, and reported by Logan (1983b).

Thus, the wet deposition of NH_4^+ to remote continental areas varies from 0.01 g to 0.47 g N/(m^2 · yr). As is the case with NO_3^-, this range is too great to make an accurate estimate of NH_4^+ in wet deposition to remote continental areas, especially given the apparent large spatial variability in the values. Therefore, until the data base improves, I have estimated that 0.08 g N/(m^2 · yr) is a reasonable value to use for global background. However, if regional estimates of the background wet deposition of NH_4^+ to remote areas are required, data from the appropriate region should be used.

8.3.5. Continental Areas: Dry Deposition

8.3.5.1. **Nitrate**. Logan (1983b) gives an estimated value for the dry-deposition rate of NO_3^- aerosol to remote continental areas of 0.006 g to 0.02 g N/(m^2 · yr). This is based on an NO_3^- concentration of 0.06 µg N/m^3 and a deposition velocity range of 0.3-1.2 cm/sec (Table 8-9). A range of 0.1-2.0 cm/sec probably comprises the real NO_3^- deposition velocity for representative earth surfaces (personal communication from

Table 8-9. Concentrations, Deposition Velocities, and Dry-deposition
Rates for Nitrogen Species in Remote Continental Areas.

Atmospheric Concentration (μg N/m^3)	Deposition Velocity (cm/sec)	Dry-Deposition Rate (g N/[m$^2 \cdot$yr])	Reference
		NO_3^-	
0.06	0.3–1.2	0.006–0.02	Logan 1983b
		HNO_3	
0.06	0.6–1.2	0.01–0.02	Logan 1983b
0.15	0.5	0.02	Böttger et al. 1978
		NO_x	
0.2	0.3–0.6	0.02–0.04	Logan 1983b
		NH_3	
0.35	0.8	0.09	Söderlund & Svensson 1976, Ayers & Gras 1983
0.0006–0.06*	0.8	0.0002–0.02	Taylor et al. 1983
		NH_4^+	
2	0.03	0.02–0.08	Söderlund & Svensson 1976
0.04–0.1	0.03–2	0.0004–0.06	Huebert & Lazrus 1980, Gravenhorst et al. 1983

*Range of NH_3 concentration in equilibrium with acidic, wet sulfate
aerosols.

Dr. G. Gravenhorst, Inst. für Meteorologie, Goethe–Universität, Frankfurt). The NO_3^- concentration in remote areas is probably on the order
of 0.03–0.3 μg N/m^3 (Gravenhorst 1975, Böttger et al. 1978, Savoie 1984).
Based on these data, dry-deposition fluxes of the order of 0.001–0.2 g
N/(m^2 · yr) are estimated for NO_3^- aerosol. For purposes of comparison
with other fluxes, I selected an average value for the dry deposition of
NO_3^- aerosol to be 0.01 g N/(m^2 · yr) with the realization that the
uncertainty about this value could be an order of magnitude (Fig. 8-5).

8.3.5.2. Nitric Acid. Logan's (1983b) estimate for the dry deposition
rate of HNO_3 of 0.01–0.02 g N/(m^2 · yr) is based on an atmospheric

concentration of 0.06 µg N/m^3 and a deposition velocity of 0.6–1.2 cm/sec (Table 8–9). Böttger et al. (1978) calculate a rate of 0.02 g $N/(m^2 \cdot yr)$ based on an atmospheric concentration of 0.15 µg N/m^3 and a deposition velocity of 0.5 cm/sec. A value of 0.02 g $N/(m^2 \cdot yr)$ for the dry-deposition flux of HNO_3 was chosen for comparison purposes.

8.3.5.3. Nitrogen Oxides.

Using an atmospheric concentration of 0.2 µg N/m^3 and a deposition velocity ranging from 0.3–0.6 cm/sec, Logan (1983b) calculates a dry-deposition rate for NO_x to remote continental areas of 0.02–0.04 g $N/(m^2 \cdot yr)$ (Table 8–9). An average flux of 0.03 g $N/(m^2 \cdot yr)$ was chosen for comparison purposes.

8.3.5.4. Ammonia.

Ayers and Gras (1983) and Söderlund and Svensson (1976) estimate that the background concentration of NH_3 in air originating over the southern Australian continent is 0.35 µg N/m^3. This is substantially higher than the estimate of Taylor et al. (1983) for the range of NH_3 concentrations (0.0006–0.06 µg N/m^3) to be expected if the concentration is controlled by equilibrium with acidic, wet $SO_4^=$ aerosol. Using a deposition velocity of 0.8 cm^2/sec (Söderlund and Svensson 1976), the dry-deposition rates calculated from these two concentration estimates are 0.09 g and 0.0002–0.02 g $N/(m^2 \cdot yr)$ (Table 8–9).

For comparison purposes I selected a value of 0.01 g $N/(m^2 \cdot yr)$ as the dry-deposition rate of NH_3 to remote continental areas while realizing that this number is very uncertain. In fact, because of the lack of experimental data, the direction in which NH_3 is being transferred between atmosphere and vegetation is unclear as is the magnitude such a flux may be having (personal communication from Dr. G. Gravenhorst, Inst. für Meteorologie, Goethe–Universität, Frankfurt).

8.3.5.5. Ammonium.

Estimates of the dry deposition of NH_4^+ depend on estimates of the dry-deposition velocity and NH_4^+ concentrations—the extremes are 0.03 cm/sec (Söderlund and Svensson 1976) and 0.5–2 cm/sec deduced from field measurements in forest canopies (Gravenhorst et al. 1983). Söderlund and Svensson (1976) estimate that the rate of dry deposition of NH_4^+ is 0.02–0.08 g $N/(m^2 \cdot yr)$, which is based on an atmospheric concentration of 2 µg N/m^3 and a deposition velocity of 0.03 cm/sec (Table 8–9). Their estimate of the NH_4^+, however, is probably influenced by anthropogenic sources. A more realistic estimate of the concentration range would probably be 0.04–0.1 µg N/m^3 (Huebert and Lazrus 1980). Therefore, using the range of deposition velocities (0.03–2 cm/sec) and the NH_4^+ concentration from Huebert and Lazrus (1980) produces a dry-deposition rate of 0.0004–0.06 g $N/(m^2 \cdot yr)$.

For comparison purposes, I selected a value of 0.001 g $N/(m^2 \cdot yr)$, again with the realization that the uncertainty of this estimate is quite large (Fig. 8–5).

8.3.6. Polar Areas

Our knowledge about the deposition of N species to polar areas is limited to a few determinations of NH_4^+ and NO_3^-. The following sections summarize these studies and present the best estimates for the deposition of NH_4^+ and NO_3^- to polar areas.

Table 8-10. Wet Deposition of NH_4^+ and NO_3^- to Polar Areas.

Location	Precipitation Accumulation Rate (cm/yr)	Concentration (μeq/l)	Deposition (g N/[$m^2 \cdot$yr])	Reference
		NH_4^+		
Greenland				
Dye 3	50	0.7	0.0049	Busenberg & Langway 1979
Antarctica				
South Pole	8.0	1.6	0.0037	Laird et al. 1982
	12	1.0	0.0017	Parker et al. 1978
	12	0.2	0.0003	Legrand et al. 1984
		NO_3^-		
Greenland				
North Central	13	1.34	0.0024	Herron 1982
Crete	27	1	0.0038	Risbo et al. 1981
Camp Century	32	1.0	0.0045	Herron 1982
Dye 3	50	0.8	0.0056	Herron 1982
Antarctica				
Vostok	2.2	0.3	0.0005	Herron 1982
Dome C	3.7	0.2	0.0003	Delmas et al. 1982a, Legrand & Delmas 1985
South Pole	8.0	1.4	0.0024	Parker et al. 1978, Legrand & Delmas 1985
D 55	8.3	0.5	0.0006	Legrand & Delmas 1985
Ross Ice Shelf	10	0.7	0.001	Herron 1982
Byrd Station	16	0.6	0.0013	Herron 1982
D 80	23	0.9	0.0029	Legrand & Delmas 1985
D 57	27	0.7	0.0026	Legrand & Delmas 1985
Siple	40	0.5	0.0028	Herron 1982
J. Ross Island	60	0.4	0.0034	Aristarain et al. 1982

8.3.6.1. Ammonium. The limited data on NH_4^+ in polar wet deposition
(Table 8-10) indicates that the values are on the order of 0.5 μeq/l and
quite variable. Busenberg and Langway (1979) state that for southern
Greenland, NH_4^+ levels range from 0.18-1.5 μeq/l with a mean and standard
deviation of 0.7 ± 0.5 μeq/l. In Antarctica, Parker et al. (1978) report
that the NH_4^+ levels are also variable (1.0 ± 1.6 μeq/l) and state that
there is a good correlation with annual sunspot activity, implying a
greater loss of ammonia by oxidation during periods of high sunspot
activity. Busenberg and Langway (1979) also show a sig-nificant correlation of NH_4^+ concentrations in Greenland with sunspot activity but caution that more data are required to confirm the relationship (Legrand et
al. 1984).

Using these estimates of NH_4^+ concentrations and the stated annual accumulation rates provide a deposition rate of NH_4^+ ranging from 0.0003 g to 0.0049 g $N/(m^2 \cdot yr)$ in polar areas (Table 8-10). For purposes of comparison to other fluxes, I selected 0.001 g $N/(m^2 \cdot yr)$.

8.3.6.2. **Nitrate**. The estimates of NO_3^- in deposition in polar areas, which are more numerous than for NH_4^+, range from 0.2-1.34 µeq/l with deposition rates ranging from 0.0003-0.0056 g $N/(m^2 \cdot yr)$ (Table 8-10). A variety of processes may be causing the observed variability of NO_3^- in polar wet deposition: supernovae (Risbo et al. 1981), solar activity (Parker et al. 1978, Laird et al. 1982), air-mass source areas (Arista-rain et al. 1982), and such meteorological factors as snowfall accumulation rates (Herron 1982). Without more data it would be difficult to determine the relative importance of the processes.

For purposes of comparison, I have selected an average value for the deposition of NO_3^- to polar areas of 0.001 g $N/(m^2 \cdot yr)$.

8.3.7. Summary

The estimates for the wet and dry deposition of N species to remote marine and continental are summarized in Figure 8-5. The following points are made from these data and the data from remote polar areas.

1. In remote marine areas, the wet-deposition rates for NH_4^+ and NO_3^- are approximately equal to or several times greater than the rates of dry deposition for either the oxidized-N compounds or the reduced-N compounds. Although the estimates of the rates of dry deposition for all N species vary widely, it is unlikely that dry deposition is as important a sink for N species as wet deposition.

2. In remote continental areas, the wet deposition of NO_3^- is about equal to the sum of the dry deposition of NO_3^-, HNO_3, and NO_x. For NH_4^+, however, the wet deposition is several times greater than the dry deposition of NH_3 plus NH_4^+ aerosol. For the N species as a whole, the estimated wet-deposition rate is about double that of the dry deposition of N species considered here whether the species is oxidized or reduced.

3. In polar areas, the rates of wet deposition of NO_3^- and NH_4^+ are approximately equal. There are no estimates of dry deposition rates available.

4. In remote marine areas, the deposition of NO_3^- and NH_4^+ is essentially equal to the deposition to remote continental areas, given the uncertainties involved in ascertaining deposition rates.

5. In polar areas, the deposition of NO_3^- and NH_4^+ is about an order of magnitude less than the deposition to remote marine and continental areas.

8.4.   REFERENCES

Andreae, M. O. 1982. Marine aerosol chemistry at Cape Grim, Tasmania, and
    Townsville, Queensland. J. Geophys. Res. 87:8875-8885.
Aristarain, A. J., R. J. Delmas, and M. Briat. 1982. Snow chemistry on
    James Ross Island (Antarctic peninsula). J. Geophys. Res. 87:
    11,004-11,012.
Ayers, G. P., and J. L. Gras. 1983. The concentration of ammonia in
    southern ocean air. J. Geophys. Res. 88:10,655-10,659.
Bonsang, B., B. C. Nguyen, A. Gaudry, and G. Lambert. 1980. Sulfate
    enrichment from marine aerosols owing to biogenic gaseous sulfur
    compounds. J. Geophys. Res. 85:7410-7416.
Böttger, A., D. H. Ehhalt, and G. Gravenhorst. 1978. Atmosphärische
    Kreisläufe von Stickoxiden und Ammoniak. Ber. der Kernforschungsan-
    lage Jülich GmbH 1558 (ISSN 0366-0885), Jülich, West Germany.
Bromfield, A. R. 1974. The deposition of sulphur in rainwater in Northern
    Nigeria. Tellus 26:408-411.
Busenberg, E., and C. C. Langway, Jr. 1979. Levels of ammonium, sulfate,
    chloride, calcium and socium in snow and ice from southern
    Greenland. J. Geophys. Res. 84:1705-1709.
Cogbill, C. V., G. E. Likens, T. A. Butler. 1984. Uncertainties in
    historical aspects of acid precipitation--Getting it straight.
    Atmos. Environ. 10:2261-2268.
Cragin, J. H., M. M. Herron, C. C. Langway, Jr., and G. Klouda. 1974.
    Interhemispheric comparison of the changes in the composition of
    atmospheric precipitation during the late Cenizoic era. In
    Conference on Polar Oceans (Montreal, May 1974) Hanover, NH:Cold
    Regions Res. and Engineering Lab.
Delmas, R. J. 1982. Antarctic sulfate budget. Nature 299:677-678.
Delmas, R., and C. Boutron. 1978. Sulfate in Antarctic snow: spatio-
    temporal distribution. Atmos. Environ. 12:723-728.
Delmas, R., and C. Boutron. 1980. Are the past variations of the
    stratospheric sulfate burden recorded in central and Antarctic snow
    and ice layers? J. Geophys. Res. 85:5645-5649.
Delmas, R., J. M. Barnola, and M. Legrand. 1982a. Gas-derived aerosol in
    central Antarctic snow and ice: The case of sulfuric and nitric
    acids. Ann. Glaciol. 3:71-76.
Delmas, R., M. Briat, and M. Legrand. 1982b. Chemistry of South Polar
    snow. J. Geophys. Res. 87:4314-4318.
Doronin, A. N. 1975. Chemical composition of snow near Vostok Station
    along the axis Mirny-Vostok. Bull. Sov. Antarct. Exp. 91:62-68.
Galloway, J. N., and A. Gaudry. 1984. The composition of precipitation on
    Amsterdam Island, Indian Ocean. Atmos. Environ. 12:2649-2656.
Galloway, J. N., and G. E. Likens. 1976. Calibration of collection
    procedures for the determination of precipitation chemistry. Water,
    Air, Soil Pollut. 6:241-258.
Galloway, J. N., and G. E. Likens. 1978. The collection of precipitation
    for chemical analysis. Tellus 30:71-78.
Galloway, J. N., G. E. Likens, W. C. Keene, and J. M. Miller. 1982. The
    composition of precipitation in remote areas of the world. J.
    Geophys. Res. 87:8771-8786.

Galloway, J. N., A. H. Knap, and T. M. Church. 1983. The composition of western Atlantic precipitation using shipboard collectors. J. Geophys. Res. 88:10,859-10,864.

Galloway, J. N., D. M. Whelpdale, and G. T. Wolff. 1984. The flux of S and N eastward from North America. Atmos. Environ. 12:2595-2607.

Georgii, H. W., and G. Gravenhorst. 1977. The ocean as source or sink of reactive trace gases. Pageoph. (Basel) 115:503-511.

Granat. L. 1976. A global atmospheric sulphur budget. In SCOPE 7: Nitrogen, Phosphorus and Sulphur--Global Cycles (B. H. Svensson and R. Söderlund, eds.), Stockholm: Ecol. Bull. 22:102-122.

Granat, L., H. Rodhe, and R. O. Hallberg. 1976. The global sulfur cycle. In SCOPE 7: Nitrogen, Phosphorus and Sulphur--Global Cycles (B. H. Svensson and R. Söderlund, eds.), Stockholm:Ecol. Bull. 22:89-134.

Gravenhorst, G. 1975. The sulfate component in aerosol samples over the North Atlantic. Meteor. Forschungsergeb 10:22-31.

Gravenhorst, G. 1978. Maritime sulfate over the North Atlantic. Atmos. Environ. 12:707-713.

Gravenhorst, G., K. P. Müller, and H. Frauhen. 1979. Inorganic nitrogen in marine aerosols. Ges. für Aerosolforschung 7:182-187.

Gravenhorst, G. K. D. Höfken, and H. W. Georgii. 1983. Acidic input to a beech and spruce forest. In Acid Deposition (S. Beilke and A. J. Elshout, eds.) Dordrecht:Reidel, 155-171.

Herron, M. M. 1982. Impurity sources of F^-, Cl^-, NO_3^-, and $SO_4^=$ in Greenland and Antarctic precipitation. J. Geophys. Res. 87:3052-3060.

Herron, M. M., C. C. Langway, Jr., H. V. Weiss, and J. H. Cragin. 1977. Atmospheric trace metals and sulfate in the Greenland ice sheet. Geochim. Cosmochim. Acta 41:915-920.

Huebert, B. J., and A. L. Lazrus. 1980. Tropospheric gas-phase and particulate nitrate measurements. J. Geophys. Res. 85:7322-7328.

Janssen-Schmidt, Th., E. P. Röth, G. Varhelyi, and G. Gravenhorst. 1981. Anthropogene Anteile am atmosphärischen Schwefel- und Stickstoffkreislauf und mögliche globale Auswirkungern auf chemische Umsetzungen in der Atmophäre. Ber. der Kernforschung Jülich GmbH 1722, Jülich, West Germany.

Keene, W. C., J. N. Galloway, A. A. P. Pszenny, and M. E. Hawley. 1985. Sea-salt corrections in precipitation chemistry. (Manuscript in preparation available from Department of Environmental Sciences, University of Virginia, Charlottesville.)

Laird, C. M. E. J. Zeller, and T. P. Armstrong. 1982. Solar activity and nitrate deposition in South Pole snow. J. Geophys. Res. Ltrs. 9:1195-1198.

Lamb, H. H. 1972. Climate. Vol. I: Past, Present, and Future. London: Methuen.

Legrand, M., and R. J. Delmas. 1984. The ionic balance of Antarctic snow, a 10-yr detailed record. Atmos. Environ. 18:1867-1874.

Legrand, M., and R. J. Delmas. 1985. Spatial variations of snow chemistry in Adelie Land (East Antarctica). (Manuscript available from authors, Rue Moliere, B.P. 96 Domaine Universitaire, St.-Martin-d'Heres, France).

Legrand, M., M. de Angelis, and R. Delmas. 1984. Ion chromatrophic
    determination of common ions at ultratrace levels in Antarctic snow
    and ice. An. Chim. Acta 156:181-192.
Likens, G. E., W. C. Keene, J. M. Miller, and J. N. Galloway. 1985. The
    chemistry of precipitation in Katherine, Australia. (Manuscript in
    preparation available from Dr. Likens, The New York Botannical
    Garden, Cary Arboretum, Millbrook, NY.)
Logan, J. A. 1983a. Comment on ''On the atmospheric input of sulfur into
    the ocean'' by E. Mészáros (1982 Tellus 34:277-282). Tellus
    35B:290-293.
Logan, J. A. 1983b. Nitrogen oxides in the troposphere: Global and
    regional budgets. J. Geophys. Res. 88:10,785-10,807.
Mészáros, E. 1982. On the atmospheric input of sulfur into the ocean.
    Tellus 34:277-282.
Més záros, E. 1983. Reply. Answer to J. A. Logan's (1983 Tellus
    35B:290-293) ''Comment on 'On the atmospheric input of sulfur into
    the ocean' by E. Mészáros.'' Tellus 35B:293.
Möller, D. 1983. The global sulfur cycle. J. Hungarian Meteorol. Serv.
    87:121-143.
Parker, B. C., L. E. Heiskell, and W. J. Thompson. 1978. Nonbiogeochemi-
    cal fixed nitrogen in anthropogenic Antarctica and some ecological
    implications. Nature 271:651-652.
Pszenny, A. A., P. F. MacIntyre, and R. A. Duce. 1982. Seasalt and the
    acidity of marine rain on the windward coast of Samoa. Geophys. Res.
    Ltrs. 9:751-754.
Risbo, T., H. B. Clausen, and K. L. Rasmussen. 1981. Supernovae and
    nitrate in the Greenland icesheet. Nature 294:637-639.
Rodhe, H., E. Mukolwe, and R. Söderlund. 1981. Chemical composition of
    precipitation in East Africa. Kenya J. Sci. and Technol. (A) 2:3-11.
Ryaboshapko, A. G. 1983. The atmospheric sulfur cycle. In The Global
    Biogeochemical Sulfur Cycle (M. V. Ivanov and G. R. Freney, eds.),
    New York:Wiley, 203-296.
Savoie, D. L. 1984. Nitrate and non-seasalt sulfate aerosols over major
    regions of the world oceans: Concentrations, sources, and fluxes.
    Ph. D. dissertation, University of Miami, Coral Gables, Florida.
Savoie, D. L., and J. M. Prospero. 1982. Particle size distribution of
    nitrate and sulfate in the marine atmosphere. Geophys. Res. Lett.
    9:1207-1210.
Sehmel, G. A. 1980. Particle and gas dry deposition: A review. Atmos.
    Environ. 14:983-1011.
Sehmel, G. A., and S. L. Sutter. 1974. Particle deposition rates on the
    water surface as a function of particle diameter and air velocity.
    J. Rech. Atm. 8:911-920.
Söderlund, R., and B. H. Svensson. 1976. The global nitrogen cycle. In
    SCOPE 7: Nitrogen, Phosphorus and Sulphur--Global Cycles (B. H.
    Svensson and R. Söderlund, eds.), Ecol. Bull. (Stockholm) 22:23-74.
Spicer, C. W. 1982. The distribution of oxidized nitrogen in urban air.
    Sci. Total Environ. 24:183-192.
Stallard, R. F., and J. M. Edmond. 1981. Geochemistry of the Amazon. 1:
    Precipitation chemistry and the marine contribution to the dissolved
    load at the time of peak discharge. J. Geophys. Res. 86:9844-9858.

Taylor, G. S., M. B. Baker, and R. J. Charlson. 1983. Heterogeneous interactions of the C, N, and S cycles in the atmosphere: The role of aerosols and clouds. In SCOPE 24: The Major Biogeochemical Cycles and Their Interactions (B. Bolin and R. B. Cook, eds.), New York: Wiley, 115-142.

Varhelyi, G., and G. Gravenhorst. 1983. Production rate of airborne sea-salt sulfur deduced from chemical analysis of marine aerosols in precipitation. J. Geophys. Res. 88:6737-6751.

Winer, A. M., J. W. Peters, J. P. Smith, and J. N. Pitts. 1974. Response of commercial chemiluninescent $NO-NO_2$ analyzers to other nitrogen-containing compounds. Environ. Sci. Technol. 8:1118-1121.

9. THE DEPOSITION OF SULFUR AND NITROGEN FROM THE REMOTE ATMOSPHERE WORKING-GROUP REPORT

Joseph M. Prospero
Rosenstiel School of Marine and
 Atmospheric Science
University of Miami
Miami, FL 33149

William C. Keene
Department of Environmental Sciences
University of Virginia
Charlottesville, VA 22903

James N. Galloway
Department of Environmental Sciences
University of Virginia
Charlottesville, VA 22903

Robert J. Delmas
Domaine Universitaire
B.P. 96, Rue Moliere
38402 St.-Martin-d'Heres, France

Lennart Granat
Department of Meteorology
University of Stockholm
Stockholm, Sweden

Gode Gravenhorst
Institut für Meteorologie und Geophysik
Johann Wolfgang Goethe-Universität
Feldbergstrasse 47
D-6000 Frankfurt a. M.1, Federal Republic of Germany

Gene E. Likens, Director
The New York Botannical Gardens
Cary Arboretum
Millbrook, NY 12545

9.1. INTRODUCTION

Materials present in the atmosphere can be deposited by a wide range of mechanisms. Historically, these deposition mechanisms have been divided

J. N. Galloway et al. (eds.), The Biogeochemical Cycling of Sulfur and Nitrogen in the Remote Atmosphere, 177–200.
© 1985 by D. Reidel Publishing Company.

into two categories: wet processes and dry processes. The wet process
most commonly studied is that involving precipitation as rain, snow, or
ice. Dry deposition to the earth's surface takes place through a wide
range of physical mechanisms including the sedimentation, impaction,
interception, and diffusion of particles and the diffusion of gases.
Another important type of deposition involves cloud and fog droplets.

In contrast to wet-deposition processes, dry-deposition processes
are very dependent on the chemical and physical character of the atmospheric constituent and of the removal surface. Consequently, it is much
easier to measure wet deposition than dry and, therefore, wet-deposition
studies are much more common than dry and the data base is much more
voluminous. For these reasons, at the Bermuda workshop we dealt with
these processes separately.

In this chapter, we discuss the various techniques used in deposition studies, assess the data available for discerning temporal and
spatial trends, discuss the causes for uncertainties in the data, and
recommend further studies.

9.2. WET DEPOSITION

9.2.1. Introduction

There are many programs for which wet deposition is collected in remote
areas and subsequently analyzed. The objectives of such studies are to
measure the wet-removal rate of atmospheric constituents; to study the
atmospheric, chemical, and physical processes controlling the cycles of
atmospheric constituents; to study the long-range transport of materials,
both natural and anthropogenic; to measure the input rates of atmospheric
constituents to ecological systems; and to determine the historical
trends in depositional patterns and, ultimately, the source strengths
through the analysis of ice cores.

9.2.2. Extent of the Data

In this assessment we have included data from two types of programs:
Those in which precipitation is being collected on an event basis and
those in which precipitation is being collected over longer periods but
where experimentation has established that the chemical composition is
not being significantly altered over the sampling period. In the absence
of such assurances, data from samples collected over longer periods can
be seriously compromised by physical, chemical, and biological processes
that could alter the chemical composition. The programs that appear to
satisfy these criteria are listed in Table 9-1. The location of these
stations is shown in Figure 9-1. Included in these listings are programs
in which the compositions of snow and ice cores are being measured.
This listing was compiled by the members of the working group on deposition using information readily at hand at the time of the workshop.
Although the list may not be comprehensive, we believe it covers all
the major programs active at the time of the Bermuda workshop.

A very extensive precipitation sampling program is being carried out
under the auspices of the WMO BAPMoN program at regional stations. Until

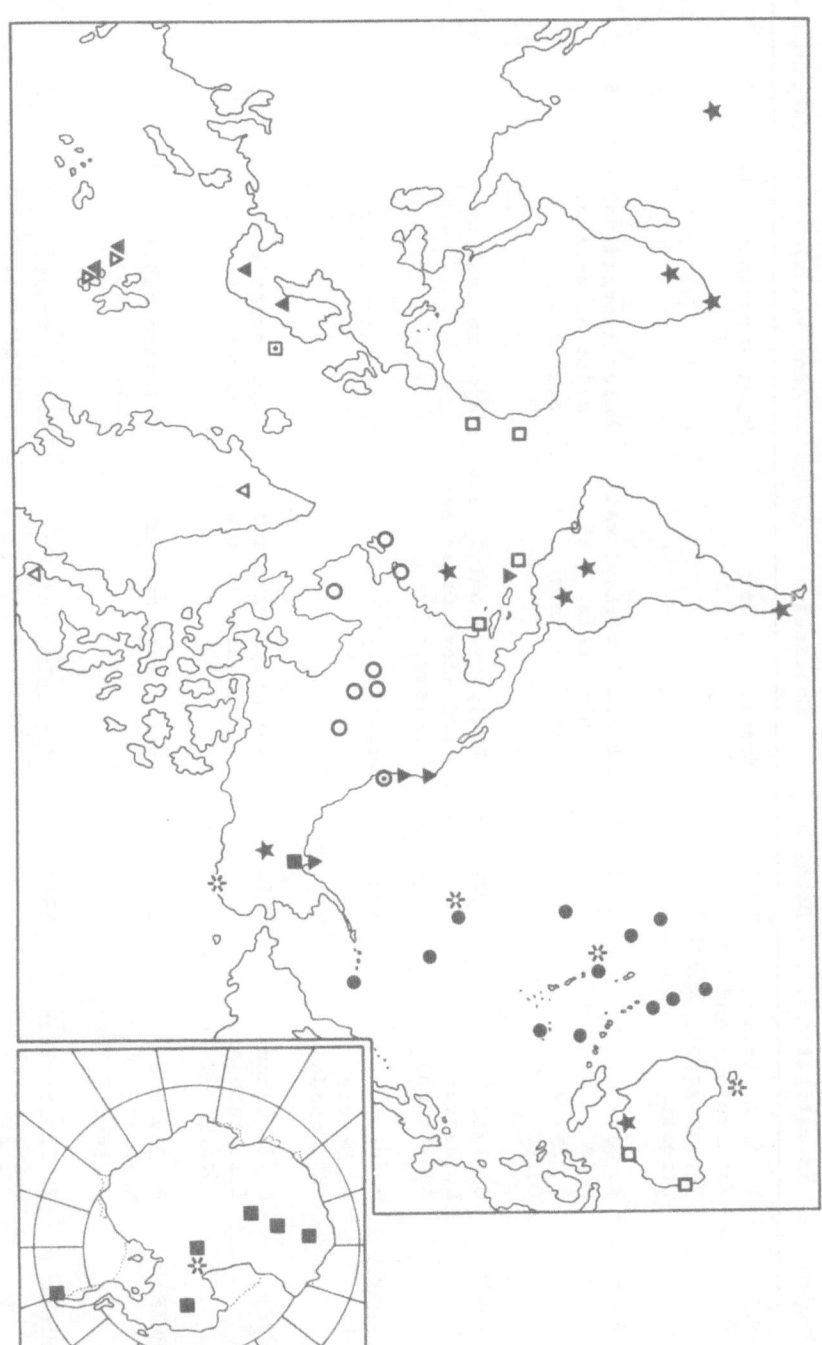

Figure 9-1. Sites of Ongoing Research Projects (see Table 8-1). □ = ASNI, ✳ = BAPMoN, ○ = CAPMON (rain and snow), △ = CAPMON (ice cores), ▼ = CWP, ▲ = EMEP, ⊙ = FIANR, ★ = GPCP, ■ = ICA, ● = SEAREX, ▽ = A. Semb and B. Holmgren, ▫ = U. of Miami.

Table 9-1. Collection and Analytical Protocols of Ongoing Research Projects (see Fig. 8-1).

Project	Principal Investigator	Beginning Date	Protocols — Collection	Protocols — Analytical	No. of Sites* — Current	No. of Sites* — Total
ASNL □	W. A. H. Asman Meteorology and Oceanography Inst. Utrecht	1978	Event, wet only	Major inorganic ions	1	1
BAPMoN ✳	R. Artz NOAA G. Ayers CSIRO	1972	Event & weekly; wet only; rain; weekly; surface snow	Major inorganics; for selected events, organics	5	5
CAPMON ○	B. Vet1 L. Barrie AES/Canada	1982	Daily; wet only; rain and snow. Gases and aerosols	Major inorganics	7	7
CAPMON △	B. Vet1 L. Barrie AES/Canada	1982	Ice cores	Major inorganics	2	2
CWP ▼	G. E. Likens Institute of Ecosystem Studies	1984	Wet only; rain and cloudwater by event	Major organics and inorganics	2	4
EMEP ▲	H. Dovland Norwegian Air Research Inst.	1974	Daily rain and snow	Selected major inorganics	2	2
FIANR ◎	R. J. Charlson University of Washington	1984	Wet only; rain by event; aerosols, gases, meteorology, cloud physics	Major, trace inorganics; major organics, aerosol optics	2	2

GPCP ★	J. N. Galloway W. C. Keene U. of Virginia	1979	Wet only; rain and snow by event	Major organics and inorganics	6	9
ICA ■	R. J. Delmas M. Legrand Glaciologie et Geophysique/Grenoble	12,000 BP	Ice cores	Major inorganics	7	7
SEAREX ●	R. Duce U. of Rhode Island J. M. Prospero U. of Miami	1981	Weekly hi-vol aerosol sampling; meteorology; some weekly; wet only; rainfall with biocide	Major organics and inorganics	5	12
▽	A. Semb Norwegian Air Research Inst. B. Holmgren U. of Upsala	19	Accumulated winter snow	Major inorganics	24 cores	11 cores
U. Miami □	J. M. Prospero U. of Miami	1973	Hi-vol aerosol sampling	Major organics	6	6

*Considered remote for the purposes of this report.

recently, the standard sampling protocol has specified month-long sampling
periods. A number of national programs are modifying (or may modify)
this protocol so that weekly samples will be collected and measures taken
to ensure the chemical integrity of the sample. We hope that more stations will follow this trend in the future.

9.2.3. Uncertainties in Measurements of Wet Deposition

Uncertainties in the estimates of wet deposition arise from a variety of
sources and they affect our ability to understand the biogeochemical
cycling of sulfur and nitrogen. Some of the more important factors
contributing to the uncertainty of deposition estimates are the analytical uncertainties, the degree the site represents the region, the dura-
tion of the data set, the collection protocol, the treatment of samples
between collection and analysis, and the uncertainties in calculating the
non-sea-salt component and in measuring the precipitation amount.

9.2.3.1. **Analytical Uncertainties.** Analytical uncertainties are a
function of accuracy, precision, and detection limits. As the number of
samples increases, uncertainties related to precision become less important in calculating average depositions at a given site. For a large
data set collected over a long period of time at a particular site, the
major uncertainties in average deposition correspond to the accuracy of
results and the detection limits of analytical techniques. For most
species, the analytical uncertainties of the measurements are small
relative to the other sources of variability.

9.2.3.2. **A Site as Representative of a Region.** Another source of uncer-
tainty derives from the natural variability of deposition processes over
space and time. This variability is an areal attribute of the environment and, therefore, does not contribute to the uncertainty of the
measurements at a particular site over a particular sampling period.
However, when an extrapolation is based on a finite data set that is
generated at a particular site, then the natural variability contributes
to the uncertainty of the extrapolation in space or in time.
      At the time of the Bermuda workshop, there were insufficient data
from remote regions to evaluate uncertainties of this nature. The only
rigorous method to quantify such uncertaintites would be to collect wet
deposition simultaneously at several sites within a specified region and
to intercompare the results based on objective criteria. This had not
yet been done.

9.2.3.3. **Duration of Data Sets.** Large-scale changes in atmospheric
circulation patterns occur on a seasonal basis in most regions of the
world. Therefore, to characterize the depositional patterns at a particular site a minimum of one year's data would be needed. Because many
data sets cover less than one year, uncertainties of this nature are,
for such sites, undefined and may be large (see Chapter 8).

9.2.3.4. **Collection Protocol.** One source of uncertainty is contamina-
tion, which can be a significant problem in the collection and analysis
of precipitation in remote regions where the concentration of dissolved

species in precipitation can be quite low. However, with careful attention to collection and handling technique, these errors can be minimized
and should not be a major cause of biased results.

If research objectives require the identification of the processes
that control the composition of wet deposition, then dry deposition must
be excluded from collected samples (Winkler 1978). In the past, it has
been a common practice to collect bulk (wet and dry) deposition samples
by using a device, such as a bucket or a funnel, that is continuously
open to the atmosphere. Such devices can give reasonable results for some
species; however, some estimates are biased because these devices typically collect an unrepresentative sample of dry deposition, both gaseous
and particulate. For best results, precipitation should be collected on
an event basis or with an automatic device that exposes the collector
only when precipitation is falling.

The sampling efficiency of wet deposition varies as a function of
the collector design, the type of precipitation, and the physical conditions, such as wind velocity. Under calm wind conditions, the uncertainties in collecting a representative sample of rainfall are small.
Under high wind conditions, collectors may preferentially sample larger
drops because smaller drops tend to follow the wind stream and be swept
past the orifice. If the chemical composition of precipitation varies as
a function of drop size (see Georgii and Wötzel 1970), this process can
result in a biased sample. Special problems occur when one attempts to
collect precipitation at sea. Because of the motion of the ship, the
often high wind speeds, and the turbulent wind conditions caused by the
hull of the vessel, quantitative collection is difficult.

The collection of snow is especially sensitive to wind effects. For
example, at exposed sites the collection efficiency of snow can be substantially less than 50%. However, with some degree of shelter from the
wind, even relatively simple devices can collect snow with an efficiency
of about 80%. (Rain collection is less sensitive to wind effects and
efficiencies of 90% or more are readily obtained.) Under strong wind
conditions when collecting snowfall on an event basis in exposed locations, snow deposited during previous events can be mixed with falling
snow and result in a biased sample. During very cold periods, sensors on
automatic collectors frequently do not detect events. In areas where
snowfall is frequent, this can yield an underestimate of wet deposition.

Cloud water and fog are important, and largely unquantified, forms
of deposition. They constitute a special case in that the controlling
deposition mechanisms are similar to those applicable to "dry" aerosol
particles. Many instruments designed to collect airborne droplets are
prototypes that have not been rigorously evaluated as to collection
efficiency (e.g., the effects of particle size). It is also very difficult to estimate the amount of water deposited during such cloud and fog
events.

9.2.3.5. **Sample Preservation.** Ideally, the chemical composition of a
precipitation sample should be measured immediately after it has been
collected. In practice, this is often impossible and, therefore, to
prevent changes in chemical composition, samples must be preserved immediately after collection. This is especially important for nutrients

used by biota, such as NH_4^+, NO_3^-, $HCOO^-$, and CH_3COO^-. Samples are typically being preserved with the addition of a biocide (Galloway and Likens 1978, Galloway et al. 1982) or by freezing (Müller et al. 1982). Such techniques should be experimentally tested to ensure that they are effective and not in themselves changing the chemical composition.

9.2.3.6. **Uncertainties in Calculating the Non-sea-salt Component.** For precipitation collected in marine areas, a major sea-salt element, such as Na^+, is often measured to calculate the fraction of the constituent derived from sea salt. Three assumptions are inherent in this procedure: First, that all of the reference element is derived from the dissolved salts in seawater; second, that no fractionation of the sea-salt components occurs during spray-droplet production; third, that no fractionation of the sea-salt component occurs after the aerosol is injected into the atmosphere or when it is scavenged by precipitation. The uncertainties in calculating the non-sea-salt component in a deposition sample arise from two major sources: the compounding of the analytical uncertainties (i.e., especially when it is necessary to take the difference between two large numbers) and the violation of the inherent assumptions.

Because of the analytical uncertainties in the measurements of both the reference element and the constituent of interest (for example, $SO_4^=$) the non-sea-salt component exhibits large uncertainties relative to the data from which it is calculated. The nature of this problem is best illustrated with an example from Amsterdam Island (Fig. 9-2): Sample A37-C was collected using a bulk collector exposed to the atmosphere for a period of 17 hr. Chemical analyses (N=3) have revealed 30.0 ± 2.4 µeq $SO_4^=$/l and 44.4 ± 2.7 µeq Mg^{++}/l. These concentrations are similar to the volume-weighted averages for the 79 events studied. Based on a seawater $SO_4^=/Mg^{++}$ ratio of 0.530, 25.5 µeq $SO_4^=$/l is contributed by sea salt and 6.5 µeq $SO_4^=$/l originates from non-sea-salt sources. Although the analytical uncertainties for the measurements of Mg^{++} and $SO_4^=$ are small (about 6%), they combine to give a total uncertainty for excess $SO_4^=$ of ± 2.8 µeq $SO_4^=$/l or ± 43% of the non-sea-salt fraction.

This example clearly demonstrates that the larger the sea-salt component is, the larger is the relative uncertainty in calculating the excess component. In continuously open collectors, dry deposition can contribute a large fraction of the total deposition of sea-salt aerosol (see Chapter 8 and Varhelyi and Gravenhorst 1983). Therefore, the amount of dry deposition in samples must be minimized, either by collecting precipitation manually on an event basis or by using an automatic collector. In the absence of such procedures, it may be difficult to obtain meaningful measurements of the excess $SO_4^=$ in remote regions where sea-salt concentrations are high and excess $SO_4^=$ concentrations are low.

Several researchers have rigorously evaluated assumptions inherent in sea-salt corrections. Data from Amsterdam Island (Keene et al. 1985) the Pacific Ocean (Prospero et al. 1985), and the Antarctic peninsula (Aristarain et al. 1982) indicate that the assumptions involved in calculating excess SO_4^{2-} in wet deposition (and aerosol) samples are apparently not valid under all conditions. The resulting uncertainties correspond to systematic bias and are not related to analytical uncertainties. In the 79 samples of wet deposition collected on Amsterdam

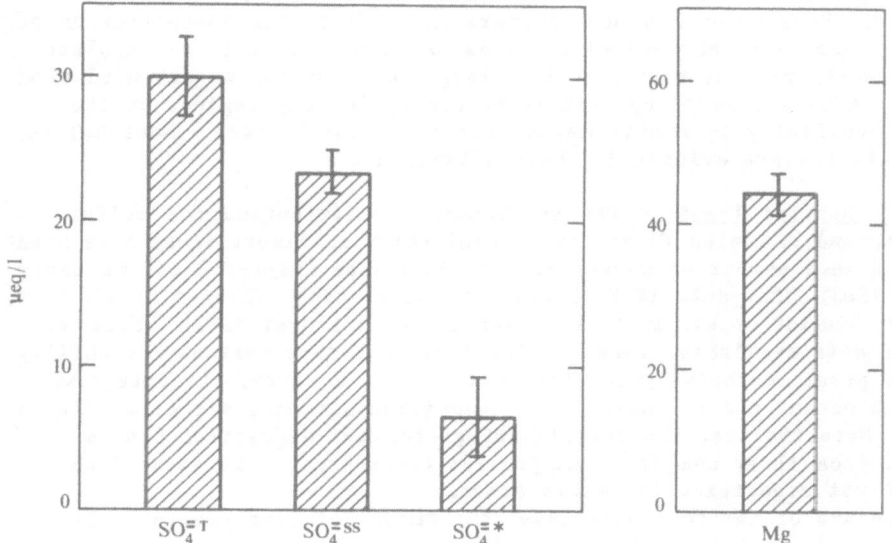

Figure 9-2. Sample A37-C From a Bulk Collector on Amsterdam Island.
            The bulk collector was exposed to the atmosphere for a period of
            17 hr.

Island, Keene et al. (1985) have found small but significant excess
concentrations of Na^+ (the most commonly used sea-salt reference element)
relative to Mg^{++}. Quality-control evaluations indicate that the excess
Na^+ is not associated with analytical errors or contamination. On a
volume-weighted basis, this corresponds to a systematic underestimate in
excess $SO_4^=$ of 14%. Consequently, the assumption that the entire concentration of airborne Na^+, Mg^{2+}, Ca^{+2} or K^+ is derived from sea salt may
not be valid; alternatively the concentration of the free ions may have
changed before chemical analyses.

9.2.4. Spatial and Temporal Trends

The spatial variability of wet deposition is a function of the distribution of the sources, the atmospheric transport patterns, the transformation processes occurring in transit, and the variability in precipitation
formation. Because of the relatively large vertical scale of precipitating systems, the composition of wet deposition tends to be more representative of the overlying atmosphere than air at the surface. Temporal
variability can occur on a wide range of time scales: minutes to hours,
because of individual precipitation-system dynamics; diurnally, because
of photochemical processes; days, because of the impact of single emission events and changing air-parcel trajectories; months, because of
seasonal changes in biological emissions and large-scale circulation;
years, because of the changes in synoptic meteorology; or decades (e.g.,

anthropogenic influence), centuries, or millenia (e.g., climatic
changes). Because of the many factors that affect the concentration of
any one species at any one site, it is often difficult to extrapolate
measurements made at that site to a larger area or for an extended time
period. A denser sampling grid is necessary in many regions of the
world, especially in remote marine and continental areas. Nonetheless,
some patterns are evident in the available data.

9.2.4.1. **Spatial Trends.** The geographical distribution for sulfur,
nitrate, and ammonium fluxes in precipitation in remote areas have been
assessed in a number of recent publications (see Chapter 8 and Galloway
et al. 1982, Whelpdale 1978, Barrie and Hales 1984, Böttger et al. 1978,
Janssen-Schmidt et al. 1981, Varhelyi and Gravenhorst 1983). From the
limited data available, these evaluations show that spatial variability
is very great, especially for the continents. However, the data from
over the oceans are so sparse that deposition patterns are difficult to
infer. Nevertheless, the overall integrated wet-deposition fluxes
derived from these compilations provide first-order estimates of the
overall wet deposition in remote areas.

   Because of the poor data base, the amount of nitrogen and sulfur
species being advected into remote areas cannot be assessed and it is
difficult to compare wet-deposition fluxes with dry-deposition fluxes and
source strengths in a defined remote region. At this time, we can only
estimate the range of wet-deposition fluxes on a global scale to demonstrate the dearth of information and the large uncertainties involved.

9.2.4.2. **Temporal Trends.** Only a few precipitation-collection sites
have operated in remote regions for more than a few years. The data from
these sites show that there is often a very large seasonal variability in
the chemical composition (see Chapter 8) and in the amount of precipitation (e.g., Likens et al. 1985).

   Longer-term trends in deposition cannot be characterized because
data records based on conventional precipitation studies are too short.
However, data from snow deposited in polar regions can be used to measure
variations in deposition for time periods ranging from months to centuries and millenia. The most extensive long-term record is based on the
analyses of ice cores from the Antarctic (Legrand 1985). The available
data show that preindustrial H_2SO_4 concentrations in Greenland (Risbo et
al. 1981, Herron 1982) and in Antarctica (Legrand and Delmas 1984) are
similar. Nitric acid concentrations are more variable in time and space.
Seasonal variations with well-pronounced maxima in summer are frequently,
but not always, observed for both acids (Aristarain et al. 1982). The
origin of HNO_3 in the polar troposphere is not known with certainty. The
sulfate apparently is derived from tropospheric transport. Stratospheric
sulfate is a minor component except after large explosive volcanic eruptions that can significantly increase stratospheric concentrations for
periods of one to two years.

   Clearly, ice-sheet studies can provide important information about
long-term trends. However, such studies do not provide a great deal of
information about deposition fluxes. The deposition rate in these
regions is very low and constitutes only a minor fraction of the global
flux of sulfur and nitrogen (see Chapter 8). In addition, the processes

controlling the composition of precipitation in polar regions are expected to be quite different from those in more moderate climates.

9.3. DRY DEPOSITION

9.3.1. Introduction

Theoretically, there is a linear relationship between the dry-deposition
flux of particles and gases to a surface and the concentrations at some
height above the surface:

$$F = V_j \cdot C_z, \qquad\qquad (9-1)$$

where F is the flux, V_j is a proportionality constant commonly called the
deposition velocity, and C_z is the concentration at some height above the
surface. However, V_j is not a simple constant; in fact, it incorporates
the eddy-transfer resistance that depends in a complex way on many surface and micrometeorological properties. For aerosols, V_j depends on
particle size, the turbulence field in the atmosphere, and the physical
structure of the surface (e.g., the roughness, the height and density of
vegetation, the dimensions of individual vegetative elements) (Friedlander and Johnstone 1957, Davidson and Friedlander 1978). Even for a
given surface and a known aerosol size distribution, large differences
are observed for the numerical values of V_j, particularly for submicron
particles. Many of these discrepancies are attributable to characteristics of the surface that are difficult to quantify, for example, its
stickiness.

The deposition of gases can involve processes, such as dissolution,
adsorption, and chemical reaction at the surface (Galbally 1974). Under
certain conditions, gas uptake by vegetation can occur by deposition on
outer surfaces and by diffusion through the stomata to the interior of
leaves and needles where biological processes provide the ultimate sink.
Deposition on the outside surfaces of plant materials can depend on the
biological utilization by microorganisms and on the presence of liquid
water on the surface. Furthermore, for some sulfur and nitrogen species,
there could be significant sources as well as sinks within a relatively
small area. Therefore, the deposition flux depends not only on the
physical and chemical properties of transfer in the lower atmosphere
(Chamberlain 1966) and at the surface but also on the biological activity
of the vegetation and in the soil (Fowler and Cape 1984, see also
Chapter 3).

Because of biological activity, the flux is not always linearly
related to concentration, contrary to what is generally assumed. For
some trace gases (NH_3, NO_2, N_2O, CO, CO_2, H_2, O_2, SO_2) (Galbally 1974,
Seiler and Giehl 1977) and for a particular concentration over a given
surface, V_j can assume a value near zero and a further decrease in
concentration can result in emission rather than deposition. The concentration at which neither deposition or emission occurs is referred to as
the compensation point. The value of the compensation point depends on
different environmental factors. However, for some species values range
between those concentrations found in industrialized areas (e.g., in

eastern North America) and those representative of remote areas. Therefore, values of V_j obtained from measurements in typical polluted continental regions may not be directly applicable to remote areas where gas
concentrations are usually much lower. In these latter regions, the
biological environment may also differ considerably and physical parameters, such as temperature, relative humidity, and radiation, are probably different.

9.3.2. Experimental Methods to Measure Dry Deposition

In theory, the deposition flux (and, in some cases, the emission flux) of
an atmospheric constituent can be measured by several techniques: (1)
collection on surrogate surfaces, (2) chamber methods, (3) micrometeorological methods, (4) ecosystem budgets, (5) atmospheric mass-balance
techniques, and (6) field-tracer studies. In principal, these techniques
can be generally applied to studies of both gaseous- and particle-
deposition fluxes. However, in practice, such experiments differ considerably in the way they are being carried out. As a consequence,
particle and gas fluxes are seldom measured simultaneously.

9.3.2.1. **Collection on Surrogate Surfaces**. Historically, most measurements of the dry deposition of particles have been made on surrogate
surfaces, such as buckets or flat plates. Such measurements are questionable because the collectors do not necessarily resemble the earth/
atmosphere interface in a chemical, physical, or micrometeorological
sense. Comparisons of various surrogate surfaces can yield widely divergent results. At one site, the fluxes of $SO_4^=$, NO_3^-, and other species
are determined simultaneously with buckets and plates over a 9-month
period and the results differ by factors of 2 to 6 (unpublished data
available from Dr. G. Gravenhorst, Goethe University, Feldbergstrasse 47,
D-6000 Frankfurt). Normally, such measurements are made with the assumption that the measured flux is totally attributable to particles: however, for some species, an unknown amount of gas-phase deposition is also
taking place. Because of the great uncertainties inherent in this technique, the data obtained are suspect and should be used with great caution, if at all.
      A variation on the surrogate method is used by Little and Wiffen
(1978) and involves trays of real grass, grown under controlled conditions. After exposure of the grass at a site, chemical analysis reveals
the amount of lead deposited on the grass and thus gives a measure of the
lead deposition rate at that site.

9.3.2.2. **Chamber Methods**. In this method, which is widely used to study
both gas and particle deposition (and emission), a chamber or wind tunnel
is used to encapsulate the study area or subject. In the case of a
ground or water surface, the chamber is simply placed over the surface--
plants and trees can be partially or totally encapsulated. The deposition of a species is determined by measuring the concentration in the
chamber as a function of biological or environmental parameters or both.
      Chamber experiments are useful for measuring the characteristic
deposition (or emission) rate of a surface in the absence of micrometeorological factors. However, chamber experiments also suffer from

various problems. They cannot simulate the micrometeorological environment. In the case of gases, reaction can occur at chamber surfaces. In many situations, the relative humidity in the chamber is difficult to control and, if condensation occurs at surfaces, the deposition fluxes can be greatly altered. A realistic photochemical environment is also difficult to maintain, especially for reactive species that are produced in the atmosphere and that have a strong surface sink. A chamber can also disturb the surface and alter the flux.

Chambers are widely used to study the deposition (and emission) rates of a wide range of sulfur and nitrogen species as well as other gases (Hällgren et al. 1982, Garland and Branson 1977, Seiler and Giehl 1977, Galbally and Roy 1978, 1980). The results from these chamber studies play an important role in atmospheric-deposition and emission studies. However, only those chamber data that include quantitative studies of the effect of the possible perturbing factors on the flux should be considered valid.

9.3.2.3. Micrometeorological Methods: Gradient and Eddy Correlation.
In the gradient method, a net vertical flux is computed as the product of the difference in the mixing ratios of a component between two levels in the atmosphere and the atmospheric turbulent-diffusion coefficient. Typically, these levels are spaced on the order of meters to tens of meters above the surface. The turbulent-eddy-diffusion coefficient for this layer can be evaluated by means of temperature and wind profiles measured simultaneously at the same site (Roth 1975).

In the eddy-correlation method, the flux is computed by relating very short-term variations in concentration with fluctuations in the vertical wind speed on a time scale ranging from minutes to a fraction of a second. Measurements are usually made at heights ranging from meters to hundreds of meters above the ground. This technique could conceivably be used in aircraft studies provided that a suitable measurement technique were available for the species of interest. However, for the vast majority of species of interest such instrumentation is not available.

These techniques require that the sources and sinks be uniformly distributed and that the atmospheric transport processes and turbulence field be homogeneous in the horizontal plane. If these conditions are met, then both methods will yield integrated fluxes that are valid for large areas. Unfortunately, the effects of inhomogeneous surface features (hedges, lines or groups of trees, bushes, hills, rivers, etc.) on the derived deposition fluxes are difficult to assess. Also, these techniques cannot be used in forests. Most studies have been made in laboratories or in the field over very smooth surfaces and under selected turbulence conditions.

Although both techniques have been applied to the study of particle deposition fluxes, the eddy-correlation method is only marginally suitable for such studies. Both methods are used for gaseous deposition studies, mainly for SO_2 and O_3 (Galbally 1974, Garland 1977, Galbally et al. 1979, Lenschow et al. 1980, Wesely et al. 1983, Garland et al. 1973). Recently, the first measurements for HNO_3 and NO_2 were reported by Wesely et al. (1982) Delany and Davies (1983) and Huebert (1983). Both methods require fairly elaborate preparation and, in the

case of the eddy-correlation technique, extensive instrumentation. The
sensitivity of both methods is only marginally either adequate or inadequate at the concentration levels characteristic of remote areas. Consequently, these techniques are not suitable for long-term deposition
studies in remote areas.

9.3.2.4. **Ecosystem Budgets.** The ecosystem-budget method does not require a knowledge of the transfer function that relates atmospheric
concentrations to the flux to the earth's surface. Instead the net
transfer across a natural interface of an ecosystem is computed by bal-
ancing input, output, and accumulation within the system. The dry deposition into an ecosystem can be determined as the flux necessary to
balance the inputs via rain against the accumulation within the system
and the output by runoff, ground-water formation, and volatilization.
The problem, however, is that the dry-deposition flux is often the result
of small differences among large and sometimes quite uncertain transfer
rates. The advantage of the budget approach is that the derived flux is
based on interactions in the real world and is not dependent on assumptions about physical processes in the boundary layer and the efficiency
of surrogate-collector surfaces.

Some important species in the ecosystem (e.g., $SO_4^=$, NH_4^+, and NO_3^-)
can be derived from the deposition of gaseous precursors (e.g., SO_2, NH_3,
and NO_x) as well as from particulate dry deposition. Furthermore, they
can be transformed to a variety of chemical and physical forms within the
system by chemical and biological processes. For such species, it can be
difficult to deduce separate gaseous and particulate inputs for each
species. Some ecosystem budgets have been computed--e.g., for the Hubbard
Brook ecosystem (Likens et al. 1977), for forest sites in the Solling
area (Ulrich et al. 1979, Ulrich and Pankrath 1983), for a spruce and
beech canopy (Gravenhorst et al. 1983), and for a desert ecosystem (Skujins 1981). However, the dry-deposition flux from the atmosphere is
often neglected in ecosystem mass balances because it is assumed that the
dry flux is small compared to other transfer rates (see e.g., contributions in Clark and Rosswall 1981).

9.3.2.5. **Atmospheric Mass-balance Techniques.** This approach is based on
the conservation of mass in a steady atmospheric flow. The divergence in
the horizontal flux (advection) of the species under consideration (in
the absence of other sources and sinks) is equal to the flux of the
species across the lower boundary, the earth's surface. Lenschow (1984)
gives an account of such a system in a more generalized form. This
technique is relatively new. It requires an extensive meteorological
data set and is generally not applicable for use in remote areas.

9.3.2.6. **Field-tracer Studies.** Field-tracer studies use isotopically
labelled species to trace the movement of materials in the environment
and to determine the rates of various transformations. Garland and his
colleagues (Garland et al. 1973, Garland and Branson 1977) have used the
isotopic species $^{35}SO_2$ to determine the deposition velocity of SO_2 to
pine forests and grasslands. In the latter case, the radioactive tracer
measurements of deposition velocity are compared to the gradient measurements of the flux and the deposition velocity. The results obtained by

the two methods agree well to within the scatter of the observations.
Similarly Rogers and his colleagues (1979) report using $^{15}NO_2$ to study
the uptake of NO_2 by vegetation. Their study shows that virtually all of
the $^{15}NO_2$ taken up by the plant is metabolized. The deposition velocity
is not explicitly determined in their experiment but, in principle, the
same technique could be used for such determinations.

The tracer technique is useful in that the source strength can be
unambiguously controlled and the mechanism of uptake can be explored in
considerable detail. Nonetheless, this technique is not widely used,
primarily because of the difficulties of working with radionuclides or,
in the case of stable isotopes, the need for stable-isotope facilities.

9.3.3. Extent of Data

Data on the dry-deposition flux of ambient airborne nitrogen and sulfur
species to the real earth's surface are very sparse and are still quite
controversial. Therefore, no values are given here for dry-deposition
velocities since they would only be relevant to the specific conditions under which they were obtained. An extensive discussion of some
problems involved and of the wide range of experimental and model results
can be found in Pruppacher et al. (1983) and Hicks (1984). Nonetheless,
for S and N species, a general picture is evident.

Of the gaseous sulfur compounds, such as SO_2, COS, CS_2, $(CH_3)_2S$,
H_2S, only the flux of SO_2 can be approximated by a deposition velocity.
Its value usually ranges from 0.1 cm/sec to 2 cm/sec (Garland 1977, Jonas
1984). The lower values are characteristic of acidic and smooth surfaces
and of quite stable atmospheric conditions whereas the higher values are
characteristic of absorbing surfaces and very rapid SO_2 transport to the
surface.

The flux of gaseous nitrogen species between the atmosphere and the
earth's surface cannot be approximated by a generalized deposition velocity. NO is usually volatilized from the soil. Above an elevated-NO-
threshold concentration in the air, the soil takes up NO. Although, NO_2
is usually absorbed from the atmosphere by soil/vegetation systems in
industrialized areas, there are indications that it may also be emitted
from natural ecosystems at lower NO_2-air concentrations. There is also
evidence that NH_3 can be volatilized as well as absorbed under conditions
prevalent at remote sites. However, no techniques are available that can
measure the direction of an NH_3 transfer or even a transfer rate. Gaseous HNO_3 is probably absorbed at the earth's surface under all conditions. It can be assumed that the HNO_3 removal rate can be approximated
by a deposition velocity similar to that for SO_2.

Particulate dry deposition is also difficult to evaluate. Some
types of areas (e.g., bare soils or ecosystems with sparse vegetation
cover) can be sources of particulate matter in the atmosphere rather
than sinks. In those areas where there is a net flux to the ground, the
deposition velocity can vary by at least an order of magnitude as a
function of particle size and the physical character of the surface. In
general, particle deposition velocities measured in the field (e.g.,
Gravenhorst and Böttger 1983) tend to be higher than those calculated
or measured in the laboratory. At high wind speeds over nonhomogeneous

terrain with permeable obstacles, the deposition velocities of particulate sulfate and ammonium can be on the order of several cm/sec. In
contrast, over smooth surfaces at low wind speed, velocities are on the
order of 0.1 cm/sec. Nitrate in remote areas is probably attached to
supermicron particles (Gravenhorst et al. 1979, Savoie and Prospero,
1982) so that NO_3^- deposition velocities can be considerably larger than
those for sulfate and ammonium in the submicron-size range.

In summary, at the time of the Bermuda workshop, we could not give
general quantitative data on dry-deposition velocities. There are still
many problems associated with the measurement of dry-deposition fluxes to
natural ecosystems because of the extremely complex chemical and physical
characters of the atmosphere/earth interface. It seemed unlikely to us
that this process would be readily quantified in the near future.

9.4. SIGNIFICANCE OF ATMOSPHERIC DEPOSITION

Measurements of atmospheric deposition provide necessary and valuable
information about the entire atmospheric cycle—emission, transport, and
transformation as well as deposition. They also provide important information about other aspects of the total biogeochemical cycle, e.g.,
ecosystem functioning.

In regard to the atmospheric cycle, data derived from the collection
and analyses of atmospheric deposition can be used to determine

- changes in temporal and spatial distributions of atmospheric
  species,
- magnitude and direction of atmospheric transport,
- reliability of atmospheric emission estimates,
- types and relative rates of transformation reactions in the
  atmosphere,
- rates of removal from the atmosphere.

The composition of wet deposition reflects the composition of the
atmosphere through which it falls. Therefore, the collection of wet
deposition is a useful tool to discern the temporal and spatial trends in
atmospheric composition. Similarly, samples of atmospheric deposition
can be used to determine the degree of transport through the atmosphere
(Galloway and Gaudry 1984, Uematsu et al. 1983, Prospero 1981). The flux
of atmospheric deposition can be used to set upper limits on the amount
of material emitted to the atmosphere. For example, in some remote
marine areas the estimated flux of $(CH_3)_2S$ from the ocean to the atmosphere closely matches the flux of excess $SO_4^=$ from the atmosphere to the
ocean. This supports the hypothesis that oceanic $(CH_3)_2S$ is the source
for most of the excess $SO_4^=$ in these regions (see Chapter 10).

The importance of some mechanisms for chemical transformation can be
validated by measuring reactant and product species in atmospheric deposition. This information can be used to set boundary conditions on the
atmospheric transformation or production rates of the species. Con-
versely, the discovery of new species in atmospheric deposition can point
to previously undescribed reactants or transformation reactions that
could be important to atmospheric chemistry. A case in point is the

discovery of HCO_2H and CH_3CO_2H in precipitation from remote areas (Keene et al. 1983). This discovery led to numerous hypotheses of transformation mechanisms that could explain the observed concentrations of these acids (Chameides 1984, Chameides and Davis 1983, Graedel and Goldberg 1983).

The above points address how measurements of atmospheric deposition can be used to achieve a better understanding of atmospheric cycling. There is, however, an equally important need for data on atmospheric deposition because of its possible impact on the functioning of terrestrial and aquatic ecosystems.

Several possible chemical inputs to remote areas could be having an impact on natural ecosystems, including nutrient (N, S, P) enrichment, geochemical weathering, trophic response, and acidification.

The atmospheric deposition of S and N in the North Temperate zone is several times larger (for similar amounts of precipitation) than for remote regions (e.g., Galloway et al. 1982). These large inputs are known to be ecologically significant for terrestrial and freshwater ecosystems. Even though the deposition rate to remote regions is appreciably smaller for most species, the evidence indicates that, for some ecosystems, these inputs constitute an important nutrient source.

Of particular importance in this regard is the deposition of nitrogen. As stated in this report, there is a significant, albeit limited, data base for NO_3^- and NH_4^+ in wet deposition from remote areas; in contrast, measurements of organic nitrogen are extremely rare. Based on these limited data, however, organic nitrogen apparently constitutes a large fraction of the total nitrogen deposited to remote areas. For example, measurements in Uganda (Visser 1961) and in a subtropical Eucalyptus forest in Australia (unpublished data available from Dr. W. Westman, Department of Geography, University of California at Los Angeles) yield organic nitrogen values that exceed those of either NO_3-N or NH_4-N.

In addition to atmospheric deposition, sources of nitrogen for biota in terrestrial ecosystems include N-fixation and weathering of minerals. Tropical rain forests, however, thrive over extensive areas of highly acidic and nutrient-poor soils. For such systems, the depositional flux of 0.24 g N/(m^2 · yr) (see Fig. 8-5) represents a major nutrient source. The vegetation in rain forests has evolved in such a manner that it recovers nutrients from wet deposition with great efficiency as water passes through the canopy and upper soil horizons (Herrera et al. 1978, Jordon et al. 1980).

Atmospheric deposition has long been recognized as an important nutrient source for surface waters of the open ocean (see review by Spencer 1975). The depositional flux of nitrogen to the sea surface is 0.08 g N/(m^2 · yr) (see Fig. 8-5). In comparison, this flux is approximately an order of magnitude greater than the 0.01 g N/(m^2 · yr) estimated for the rate of N-fixation by pelagic marine plankton (Capone and Carpenter 1982). Although the data are sparse, it appears that, for oceanic regions remote from major areas of upwelling and from riverine inputs, atmospheric deposition is the major source of nitrogen for the support of marine primary productivity.

9.5. CONCLUSIONS AND RECOMMENDATIONS

9.5.1. Wet Deposition

9.5.1.1. Spatial and Temporal Representativeness. When we met in Bermuda, the data base was inadequate to serve as the basis for designing a global deposition network. As an initial minimal configuration, such a network should consist of at least one station in each of the major wind regimes in each of the continental and ocean regions. The data base could be broadened dramatically if some existing network, such as BAPMoN, were to adhere to a suitable deposition-sampling protocol.

Long-term measurements are critically needed for the Southern Hemisphere. Such measurements would serve to establish background values and, over the longer term, to determine trends that could be related to anthropogenic impacts. In some ocean regions, no islands are suitably located to serve as sampling stations for some major wind regimes. In such cases, efforts should be made to establish sampling programs on ships of opportunity.

9.5.1.2. Integration With Other Atmospheric Cycling Measurements. The deposition process is but one aspect of the entire atmospheric cycle of S and N. Deposition data become more meaningful if they are accompanied by data related to these other processes—emission, transport, and transformation. Deposition-measurement programs should be planned accordingly. Important deposition-related aspects of these processes are discussed in the other parts of this book.

At this point, we can simply emphasize the need to make deposition measurements in areas representative of major source types and source regions, whether natural or anthropogenic. It is especially important that deposition measurements be made in conjunction with emission measurements.

9.5.1.3. Measurements Critical to Reducing Uncertainties.

1. Precipitation samples should be collected on an event basis using rain collectors that exclude dry deposition. Precautions should be taken to maintain the chemical integrity of the samples.

2. As a start, long-term collection sites for wet deposition and atmospheric gases and aerosols should be established in each of the 10 major environmental zones.

3. New types of shipboard collectors must be designed and tested.

4. Organic N and organic acids should be measured in precipitation from remote areas.

5. The regional representative collection sites should be evaluated in those areas where rainfall gradients are large or where localized sources are strong (e.g., the North Pacific, the North Atlantic, the Indian Ocean).

9.5.2. Dry Deposition

9.5.2.1. Integrate Measurements of Atmospheric Constituents with Wet-deposition Measurements.
The concentration of several aerosol- and gas-phase species should be measured in conjunction with wet-deposition studies. For some species, dry-deposition velocities could be estimated on the basis of such data. Aerosol data would also be useful in transport studies. These data are especially important because many meteorological events have a low-precipitation probability. It might also be desirable to sample for selected gas species, such as SO_2, using chemically treated filters.

9.5.3. Extend Existing Depositional-measurement Techniques

9.5.3.1. Isotopes.
Source-identification techniques should be developed based on the isotopic ratios of $^{34}S/^{32}S$ and $^{15}N/^{14}N$. Isotope ratios are especially good tracers because the chemical behavior of isotopes is identical--fractionation occurs because of differences in physical properties caused by the different isotopic masses. The isotopic ratios in a sample could provide information on the source of the material. For example, some of the major sources for atmospheric sulfur compounds (e.g., sea salt, volcanoes, biological H_2S) have a characteristic isotope ratio. Because of the small elemental mass difference between ^{34}S and ^{32}S and between ^{15}N and ^{14}N, the isotope ratios are not changed appreciably by reactions occurring during atmospheric transport.

Work on the development of nitrogen-isotope techniques should be encouraged. Experiments should be carried out to characterize the source areas. The isotopic composition of deposition samples from remote regions should be measured to determine if any trends can be related to sources. Although some values for samples of marine precipitation and aerosols and for stratospheric aerosols have been published, extensive surveys of both source areas and sink regions are needed.

9.5.3.2. Ecosystem Budgets.
The principal advantage of this approach is that the derived fluxes are representative of the complexity and character of the natural system being studied. However, it also has several disadvantages. It is labor intensive and expensive and the accumulated errors can be large because the computed net input is small and is the difference between two large values.

We recommend that, where possible, wet deposition be measured in conjunction with other elemental fluxes in and through terrestrial and aquatic ecosystems not only to determine the removal from one reservoir but also to determine the significance of the input to the receiving reservoir.

9.5.3.3. Other Important Chemical Constituents.
Deposition measurements should be extended to those species that are important to our understanding of the sulphur and nitrogen cycles. The existing wet-deposition data sets consist almost exclusively of measurements of the principal oxidative end members of the S and N cycles: $SO_4^=$ and NO_3^-. However, to understand the entire cycles of these elements, measurements of other

related species are needed. Generally, these measurements pose a greater challenge to our analytical capabilities. For maximum benefit, the following constituents should be measured in air and clouds as well as in rain.

1.   Dimethyl sulfoxide: This is a suggested oxidation product of $(CH_3)_2S$.

2.   Methane sulfonic acid: This measurement is very important since CH_3SO_2OH is a significant product in the oxidation of dimethyl-sulfide, at least under laboratory conditions.

3.   Hydroxy methane sulfonic acid: This species is produced as an aqueous complex of formaldehyde and SO_2. Its presence would indicate a substantial inhibition of SO_2 oxidation.

4.   Formaldehyde: This component is not only important because it acts as as an inhibitor of SO_2 oxidation but also because it significantly affects the HO radical concentrations and provides a check on the carbon cycle.

5.   Hydrogen peroxide: This compound along with O_3 is thought to be the major oxidizer of SO_2 in solution. It also plays a major role in determining HO concentrations. Organic peroxides could be almost as important and essentially nothing is known about them.

6.   Ammonia and the amines: These interact strongly with the sulfur cycle in clouds and aerosols.

7.   Organic nitrogen: These are suspected of being the major carriers of N in the remote atmosphere. Their concentration is unknown but could be comparable to the concentration of all other species of reactive and reservoir nitrogen.

8.   Acetone: This is an indicator of the recent influence of non-methane hydrocarbons on air chemistry.

9.   Formic and acetic acids: These are major sources of free acidity in precipitation from remote regions and thereby control, in part, all pH-dependent transformations of S and N species.

9.6. REFERENCES

Aristarain, A. J., R. J. Delmas, and M. Briat. 1982. Snow chemistry on James Ross Island (Antarctic peninsula). J. Geophys. Res. 87:11,004–11,012.

Barrie, L. A., and J. M. Hales. 1984. The spatial distribution of precipitation acidity and major ion wet deposition in North America during 1980. Tellus 36b:333–355.

Böttger, A., D. H. Ehhalt, and G. Gravenhorst. 1978. Atmosphärische
    Kreisläufe von Stickoxiden und Ammoniak. Ber. der Kernforschungsan-
    lage Jülich GmbH 1558 (ISSN 0366-0885), Jülich, West Germany.
Chamberlain, A. C. 1966. Transport of Gases to and from grass and grass-
    like surfaces. Proc. Roy. Soc. Ser. A 290:236
Chameides, W. L. 1984. The photochemistry of a remote marine stratiform
    cloud. J. Geophys. Res. 89:4739-4755.
Chameides, W. L., and D. D. Davis. 1983. Cloud chemistry: An aqueous-
    phase source of formic acid. Nature 304:427-429.
Clark, F. E., and T. Rosswall (eds.). 1981. Terrestrial Nitrogen Cycles:
    Processes, Ecosystem Strategies and Management Impacts. Stockholm:
    Ecol. Bull. No. 33, 714 pp.
Capone, D. G., and E. J. Carpenter. 1982. Nitrogen fixation in the marine
    environment. Science 217:1140-1142.
Davidson, C. I., and S. K. Friedlander. 1978. A filtration model for
    aerosol dry deposition: Application to trace metal deposition from
    the atmosphere. J. Geophys. Res. 83:2343-2352.
Delany, A. C., and T. D. Davies. 1983. Dry deposition of NO_x to grass in
    rural East Anglia. Atmos. Environ. 17:1391-1394.
Fowler, D., and J. N. Cape. 1984. The contamination of rain samples by
    dry deposition on rain collectors. Atmos. Environ. 18:183-189.
Friedlander, S. K., and H. F. Johnstone. 1957. Deposition of suspended
    particles from turbulent gas streams. Indus. Engin. Chem. 49:1151.
Galbally, I. E. 1971. Ozone profiles and ozone fluxes in the atmospheric
    surface layer. Q. J. Roy. Met. Soc. 97:18-29.
Galbally, I. E. 1974. Gas transfer near the earth's surface. Adv. in
    Geophys. 18B:329-339.
Galbally, I. E., and C. R. Roy. 1978. Loss of fixed nitrogen from soils
    by nitric oxide exhalation. Nature 275:734-735.
Galbally, I. E., and C. R. Roy. 1980. Destruction of ozone at the earth's
    surface. A. J. Roy. Met. Soc. 106:599-620.
Galbally, I. E., J. A. Garland, and M. G. A. Wilson. 1979. Sulphur uptake
    from the atmosphere by forest and farmland. Nature 280:49-50.
Galloway, J. N., and A. Gaudry. 1984. The composition of precipitation
    on Amsterdam Island, Indian Ocean. Atmos. Environ. 18:2649-2656.
Galloway, J. N., and G. E. Likens. 1978. The collection of precipitation
    for chemical analysis. Tellus 30:71-82.
Galloway, J. N., G. E. Likens, W. C. Keene, and J. M. Miller. 1982. The
    composition of precipitation in remote areas of the world. J.
    Geophys. Res. 87:8771-8786.
Garland, J. A. 1977. The dry deposition of sulphur dioxide to land and
    water surfaces. Proc. R. Soc. Lond. A.354:245-268.
Garland, J. A., and J. R. Branson. 1977. The deposition of sulphur di-
    oxide to pine forest assessed by a radioactive tracer method.
    Tellus 29:445-454.
Garland, J. A., W. S. Clough, and D. Fowler. 1973. Deposition of sulphur
    dioxide on grass. Nature 242:256-257.
Georgii, H. W., and D. Wötzel. 1970. On the relation between drop size
    and concentration of trace elements in rainwater. J. Geophys. Res.
    75:1727-1731.

Graedel, T. C., and K. I. Goldberg. 1983. Kinetic studies of raindrop
    chemistry: 1. Inorganic and organic processes. J. Geophys.
    Res. 88:10,865-10,882.
Gravenhorst, G., and A. Böttger. 1983. Field measurements of NO and NO$_2$
    fluxes to and from the ground. In Acid Deposition (S. Beilke and A.
    J. Elshout, eds.) Dordrecht:Reidel, 172-184.
Gravenhorst, G., K. P. Müller, and H. Franken. 1979. Inorganic nitrogen
    in marine aerosols. Ges. für Aerosolforsch. 7:182-187.
Gravenhorst, G., K. D. Höfken, and H. W. Georgii. 1983. Acidic input to
    a beech and spruce forest. In Acid Deposition (S. Beilke and A. J.
    Elshout, eds.) Dordrecht:Reidel, 155-171.
Hällgren, J-E., S. Linder, A. Richter, E. Tröng, and L. Granat. 1982.
    Uptake of SO$_2$ in shoots of Scotts pine: Field measurements of net
    flux of sulfur in relation to stomatal conductance. Plant, Cell,
    Environment 5:75-83.
Herrera, R., C. F. Jordan, H. Klinge and E. Medina. 1978. Amazon ecosys-
    tems. Their structure and functioning with particular emphasis on
    nutrients. Interciencia 3:223-231.
Herron, M. M. 1982. Impurity sources of F$^-$, Cl$^-$, NO$_3{}^-$, and SO$_4{}^{2-}$ in
    Greenland and Antarctic precipitation. J. Geophs. Res 87:3052-3060.
Hicks, B. B. (ed.). 1984. Deposition Both Wet and Dry. Vol. 4, Acid
    Precipitation Series (J. I. Peasley, ed.), Ann Arbor, MI:Ann Arbor
    Science.
Huebert, B. J. 1983. The dry deposition of HNO$_3$ vapour as a sink for NO$_y$.
    Presented at the CACGP Topospheric Chemistry Symp., Oxford, U.K.
    (paper available from Dr. Huebert, Colorado College, Colorado
    Springs).
Janssen-Schmidt, E., P. Röth, G. Varhelyi, and G. Gravenhorst. 1981.
    Anthropogene Anteile am atmosphärischen Schwefel- und Stickstoff-
    kreislauf und mögliche globale Auswirkungern auf chemische Umset-
    zungen in der Atmophäre. Ber. der Kernforschung., Jülich GmbH
    1722, Jülich, West Germany.
Jordon, C., F. Galley, J. Hall, and J. Hall. 1980. Nutrient scavenging of
    rainfall by the canopy of an Amazonian rain forest. Biotropica
    12:61-66.
Jonas, R. 1984. Ablagerung und Bindung von Luftveinureiniguugen an Vege-
    tation und anderen atmosphärischen Grenzflächen. Ber. der Kern-
    forsch. Jülich GmbH 1949 (ISSN 0366-0885), Jülich, West Germany.
Keene, W. C., J. N. Galloway, and J. D. Holden, Jr. 1983. Measurement of
    weak organic acidity in precipitation from remote areas of the world
    J. Geophys. Res. 88:5122-5130.
Keene, W. C., J. N. Galloway, A. A. Pszenny, and M. E. Hawley. 1985. Sea-
    salt corrections in precipitation chemistry. (Manuscript available
    from W. C. Keene, Department of Environmental Sciences, University
    of Virginia, Charlottesville.)
Legrand, M. 1985. Chimie des neiges et glaces Antarctiques: Un reflect de
    l'environnement. Master's Thesis. University of Grenoble, France.
Legrand, M., and R. Delmas. 1984. The ionic balance of Antarctic snows: A
    10-year detailed record. Atmos. Environ. 18:1867-1874.
Lenschow, D. 1984. Instrumentation development needs for use of mass-
    balance technique. In Global Tropospheric Chemistry: A Plan for
    Action, Washington, D.C.:National Academy Press, 141-143.

Lenschow, D. H., A. C. Delany, B. B. Stankov, and D. H. Stedman. 1980. Airborne measurements of the vertical flux of ozone in the boundary layer. Boundary Layer Meteorology 19:249-265.

Likens, G. E., F. H. Bormann, R. S. Pierce, N. M. Johnson, J. S. Eaton. 1977. Biogeochemistry of a Northern Hardwood Forest Ecosystem. New York: Springer-Verlag.

Likens, G. E., W. C. Keene, J. M. Miller, and J. N. Galloway. 1985. The chemistry of precipitation in Katherine, Australia. (Manuscript available from Dr. Likens, The New York Botannical Gardens, Cary Arboretum, Millbrook, NY.)

Little, P., and R. D. Wiffen. 1978. Emission and deposition of lead from moter exhausts: 2. Airborne concentration, particle-size, and deposition of lead near motorways. Atmos. Environ. 12:1331-1341.

Müller, K. P., G. Aheimer, and G. Gravenhorst. 1982. The influence of immediate freezing on the checmial composition of rain samples. In Deposition of Atmospheric Pollutants (H.-W. Georgii and J. Pankrath, eds.), Dordrecht:Reidel, 125-132.

Prospero, J. M. 1981. Eolian transport to the world ocean. In The Oceanic Lithosphere: The Sea (C. Emiliani, ed.), New York: Wiley 7:801-874.

Prospero, J. M., D. Savoie, R. T. Nees, R. A. Duce, and J. Merrill. 1985. Particulate sulfate and nitrate in the boundary layer over the North Pacific Ocean. J. Geophys. Res. (in press).

Pruppacher, H. R., R. G. Semonin, and W. G. N. Slinn (eds.). 1983. Precipitation Scavenging, Dry Deposition, and Resuspension. Vol. 2: Dry Deposition and Resuspension. New York: Elsevier.

Risbo, T., H. B. Clausen, and K. L. Rasmussen. 1981. Supernovae and nitrate in the Greenland Ice Sheet. Nature 294:637-639.

Rogers, H. H., J. C. Campbell, and R. J. Volk. 1979. Nitrogen-15 dioxide uptake and incorporation by Phaseolus vulgaris. Science 206:333-335.

Roth, R. 1975. Der vertihale Transport von Luftbeimengengen in der Praudte-Schicht und die Depositionsgesdewindigheit. Met. Rundshau. 28:65-71.

Savoie, D. L., and J. M. Prospero. 1982. Particle size distribution of nitrate and sulfate in the marine atmosphere. Geophys. Res. Lett. 9:1207-1210.

Seiler, W., and Giehl. 1977. Influence of plants on the atmospheric carbon monoxide. Geophs. Res. Lett. 4:329-332.

Skujins, J. 1981. Nitrogen cycling in arid ecosystems. In Terrestrial Nitrogen Cycles: Processes, Ecosystem Strategies and Management Impacts. Stockholm:Ecol. Bull. No. 33, 477-492.

Spencer, C. P. 1975. The micronutrient elements. In Chemical Oceanography (J. P. Riley and G. Skirrow, eds.), London:Academic Press, 365-385.

Uematsu, M., R. A. Duce, J. M. Prospero, L. G. Chen, J. T. Merrill, and R. L. McDonald. 1983. Transport of mineral aerosol from Asia over the North Pacific Ocean. J. Geophys. Res. 88:5343-5352.

Ulrich, B., and J. Pankrath (eds.). 1983. Accumulating Air Pollutants in Forest Ecosystems. Dordrecht:Reidel.

Ulrich, B., R. Mayer, and P. K. Khanna. 1979. Deposition von Luftverunreinigungen und ihre Auswirkungen in Waldökosystemen im Solling. Frankfurt:Sauerländer Verlag.

Varhelyi, G., and G. Gravenhorst. 1983. Production rate of airborne sea-salt sulfur deduced from chemical analysis of marine aerosols and precipitation. J. Geophys. Res. 88:6737–6751.

Visser, S. 1961. Chemical composition of rainwater in Kampala Uganda, and its relation to meteorological and topographical conditions. J. Geophys. Res. 66:3759–3765.

Wesely, M. L., J. A. Eastman, D. H. Stedman, and E. D. Yalvac. 1982. An eddy correlation measurement of NO$_2$ flux to vegetation and comparison to O$_3$ flux. Atmos. Environ. 16:815–820.

Wesely, M. L., D. R. Cook, R. L. Hart, B. B. Hicks, J. L. Durham, R. E. Speer, D. H. Stedman, R. J. Tropp. 1983. Eddy-correlation measurements of the dry deposition of particulate sulfur and submicron particles. In Precipitation Scavenging, Dry Deposition, and Resuspension, Vol 2: Dry Deposition and Resuspension (H. R. Pruppacher, R. G. Semonin, and W. G. N. Slinn, eds.), New York:Elsevier, 943–952.

Whelpdale, D. M. 1978. Atmospheric Pathways of Sulfur Compounds. MARC Rept. 7, London:Chelsea College.

Winkler, P. 1978. Fehler bei der Spurenstoffanalyse im atmosphärischen Niederschlag dargestellt am Beispiel von pH-Wert und Elektrischer Leitjähigkeit. Stanb. 38:175–177.

PART V

OVERVIEW AND SUMMARY

10. THE CYCLING OF SULFUR AND NITROGEN IN THE REMOTE ATMOSPHERE

Paul J. Crutzen, Director
Airchemistry Department
Max-Planck-Institut für Chemie
P. O. Box 3060
D-6500 Mainz, Federal Republic of Germany

Douglas M. Whelpdale
Environment Canada
Atmospheric Environment Service
4905 Dufferin Street
Downsview, Ontario
Canada M3H5T4

Dieter Kley*
Atmospheric Sampling Group
Aeronomy Laboratories/NOAA
325 Broadway
Boulder, CO 80303

Leonard A. Barrie
Environment Canada
Atmospheric Environment Service
4905 Dufferin Street
Downsview, Ontario
Canada M3H5T4

10.1. INTRODUCTION

The main part of this book has described in detail the important components of the atmospheric cycles of sulfur and nitrogen--emission, transformation, transport, and deposition. Towards the end of the Bermuda
workshop, two brief sessions were convened to discuss specific aspects of
the two atmospheric cycles. These discussions focussed on parts or
combinations of parts of special concern to the participants. However,
these discussions did not result in atmospheric budgets for sulfur and
nitrogen not only because of a lack of time but also because of the large
uncertainties in the deposition, especially in the emission estimates.

---

*Present address: Institut für Chemie der Kernforschungsanlage Jü-
lich GmbH, Postfach 1913, D-5170 Jülich 1, Federal Republic of Germany.

J. N. Galloway et al. (eds.), The Biogeochemical Cycling of Sulfur and Nitrogen in the Remote Atmosphere, 203–212.
© 1985 by D. Reidel Publishing Company.

10.2. SULFUR CYCLE

10.2.1. Emission and Deposition in Marine Areas

The working group on emissions (Chapter 3) estimated that 0.11 g S/(m^2 · yr) is emitted to the marine atmosphere as DMS. Emission rates of other sulfur species (e.g., H$_2$S, COS) are unknown but thought to be insignificant for the tropospheric S cycle. Galloway (see Chapter 8) estimates that the wet- and dry-deposition rates of excess sulfur in marine areas are 0.1 g S/(m^2 · yr) and 0.04 g S/(m^2 · yr), respectively, for a total of 0.14 g S/(m^2 · yr). Given the uncertainties of the emission and deposition estimates, one cannot exclude the possibility that other sulfur compounds (e.g., H$_2$S) are also being emitted to the marine atmosphere at rates comparable to DMS. In other words, the state of our knowledge on the sources and sinks of sulfur in the remote marine atmosphere is incomplete.

We know that emission and deposition rates of sulfur exhibit significant temporal and spatial variabilities: summer and winter deposition rates differ and biological productive regions have higher sulfur-emission rates than less productive ones (see Chapter 1). However, as Table 10-1 indicates, the data for remote areas are still insufficient to determine quantitatively the spatial and temporal variabilities.

In addition to comparing emission and deposition rates within the atmospheric sulfur cycle as a check on the completeness and consistency of our knowledge (Table 10-1), the rates of transformation (see Part II) from reduced- to oxidized-sulfur species can also be examined in this manner. Andreae (1985) has illustrated the current state of knowledge of sulfur cycling in the tropical marine boundary layer--the controlling processes, the fluxes, the typical reservoir concentrations, and the lifetimes (Fig. 10-1). This combination of measurements and estimates of the reservoir flux yields a consistent picture of the cycle--from the emission of DMS to the deposition of SO$_2$, SO$_4^=$, CH$_3$SO$_3$H, and possibly other species. However, new information could alter this picture. For example, dimethylsulfide-hydroperoxide (CH$_3$SCH$_2$OOH) has not yet been discussed in marine atmospheric chemistry. Its chemistry and photochemistry are unknown but, given the uncertainties in the cycle, it could exist. This would necessitate a reevaluation of the current thinking concerning this "closed" cycle current.

10.2.2. Emission and Deposition in Continental Areas

The estimates of total emissions of SO$_2$, H$_2$S, DMS, and SO$_4$ from volcanoes, soils, plants, coastal wetlands, and biomass burning to the continental atmosphere range from 0.16 g to 0.44 g S/(m^2 · yr) with the majority (0.02-0.28 g S/[m^2 · yr]) coming from H$_2$S emissions from soils and plants (see Chapter 1). Wet deposition and dry deposition to remote continental areas are estimated to be 0.2 g S/(m^2 · yr) and 0.1 g S/(m^2 · yr) with a total deposition of 0.3 g S/(m^2 · yr) (see Chapter 8). (The uncertainty in these last value could be as much as a factor of 5.) The uncertainties indicated some of the shortcomings of our knowledge of the atmospheric sulfur cycle over remote continental areas. Given the poor agreement between the emission and deposition rates, an internal

Table 10-1. Plausible Values of Emission and Deposition Rates of S Species (g S/[$m^2 \cdot$ yr]). Deposition values are for remote areas only.

Boundary Layer	Emission				Deposition		
	H_2S[a]	DMS	SO_2[b]	Total	Wet[c]	Dry[d]	Total
Polar							
Winter	0	0	?	?	0.003[e]	?	≥0.003
					0.04		≥0.04
Summer	?	?	?	?	0.003[e]	?	≥0.003
					0.04		≥0.04
Midlatitude: Winter							
Continental	?	?	0.5[f]	≥0.5	?	0.1	>0.1
Oceanic	?	?	0	?	0.1–0.2[g]	0.04	>0.1–0.2
Midlatitude: Summer							
Continental	?	0.003	0.4[f]	≥0.4	?	0.1	≥0.1
Oceanic	?	0.1	0	≥0.1	0.1–0.2[g]	0.04	>0.1–0.2
Subtropical							
Continental	?	?	?	?	0.12–0.4	0.1	>0.2–0.5
Oceanic	?	0.1	0	≥0.1	?	0.04	>0.04
Equatorial							
Continental	?	?	?	?	0.2	0.1	0.3
Oceanic	?	0.14	0	≥0.1	?	0.04	≥0.04

[a]Includes other sulfur gases.

[b]Includes sporadic emissions from volcanoes.

[c]Based on limited measurements.

[d]Values are order-of-magnitude estimates based on limited atmospheric concentration data and assumed deposition velocities.

[e]Low value is for Greenland (~ 2,500 m); high value, for Alaska and Spitzbergen (~ 100 m).

[f]Mainly from Northern Hemisphere; includes anthropogenic emissions within polluted regions (~ 1,000 · 1,000 km) emission ≅ 3.

[g]The influence of anthropogenic emissions likely.

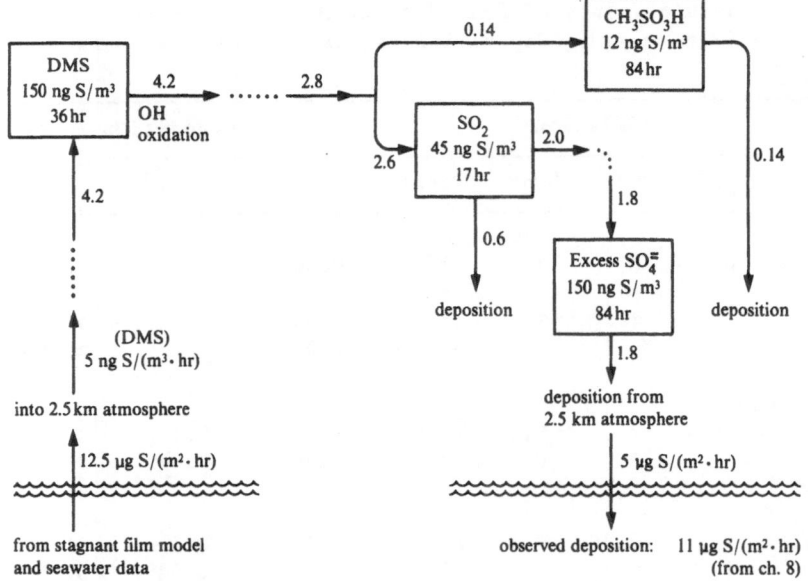

Figure 10-1. The Sulfur Cycle in the Remote Marine Troposphere.
Data on the sea-to-air flux of DMS is from Andreae and Raemdonck
(1983), on the DMS concentration and lifetime from Andreae et al.
(1985), on the methanesulfonic acid concentration from Andreae
(1985), on the concentration and residence time of SO_2 from
Bonsang et al. (1980), and on the concentration and deposition of
excess sulfate from Galloway (Chapter 8).

consistency check on emission, transformation, and deposition rates such
as was done for the marine atmosphere (Andreae 1985) could not be dupli-
cated for the remote continental atmosphere.

10.2.3. Measurement and Analysis Strategies

All gases (except COS and N_2O) actively involved in the atmospheric
sulfur and nitrogen cycles have rather short atmospheric residence times,
ranging from very short to about one week. As a result, their atmospheric abundances exhibit considerable patchiness, which is primarily
caused by meteorological processes. The design of a network of atmospheric observatories, the planning of research expeditions, and the
interpretation of chemical data should all, therefore, be aided by com-
plementary meteorological data analyses on appropriate time and space
scales (see Part III).
       Carefully selected atmospheric models could also serve as important
tools but not all modeling approaches would be appropriate. For example,
one-dimensional, eddy-diffusion models, which have been applied so suc-
cessfully to stratospheric research, would only be useful for the study
of atmospheric constituents with lifetimes longer than about one month.

However, to study S and N cycles, these models would clearly be inadequate because they do not take into account the meteorological characteristics of the planetary boundary layer. For example, they do not
account for the upward transport to the middle and upper troposphere that
short-lived S and N gases may experience at frontal zones and during
thunderstorms.

10.2.4. Continental Sources of Sulfur

Many observations indicate the existence of additional continental
sources of sulfur in the tropics and subtropics: The magnitude of the wet
deposition of S found in samples from the Ivory Coast of Africa and in
Venezuela (Galloway et al. 1982) indicate that stronger sulfur emissions
than are currently known must occur in the marine environment.

Dry deposition could be a very important additional sink mechanism
in this area of the world. For much of the year, precipitating clouds
cover only a small fraction of the region at any given time and the
strong vertical turbulent exchange to the surface, along with the rough,
moist, alkaline nature of many surfaces, would particularly favor the dry
deposition of SO_2. The cycling of alkaline sea-salt particles through
the lower marine boundary layer and their interaction with water vapor
make them potentially very efficient scavenging surfaces for acidic
gases.

These bits of evidence indicate that very effective removal processes are active in the tropics and subtropics. The need to supply as
much sulfur to the atmosphere as is thought to be removed supports the
existence of more continental sources of sulfur in the tropics and subtropics than have yet been found. This evidence also points to the need
for a much better understanding of the dry-deposition processes and
fluxes in these regions.

One potentially useful approach to these problems would be the
measurement of the isotopic ratios of S. With seawater and probably DMS
having a relatively uniform value of δS +20 per mil, significant deviations from this value in marine aerosols or wet deposition would point to
a nonmarine source for the sulfur.

10.2.5. The Need for Additional Measurement Sites

The important, distinct meteorological regimes already established around
the globe, the limited deposition data currently available, and the lack
of reliable emission data exemplify the need for additional remote continental measurement stations (wet deposition and air concentrations) in
the tropics, subtropics, and midlatitudes. We all agreed that more than
the traditional "one site in a remote area" would now be more appropriate. For example, large continental source areas could be characterized by measurements from stations strategically located in a line
inland from the coast or specified continental ecosystems by measurements
from several stations placed throughout the ecosystem (see Chapter 6). A
few stations located in a line inland from the Pacific Ocean (e.g., on
the Olympic Peninsula in Washington state or in the Andes of South
America) could be used to characterize the marine-layer/free-troposphere
interface.

Even deserts and glaciers—although extremely inhospitable environments in which to make deposition measurements—may be important sink
regions.   We have no information about the importance of dry deposition
in parts of such regions where precipitation amounts are small.

As a final point, we noted that most of our deposition information
in the marine environment comes from island sites. We need to establish
whether a bias has been or is being introduced into the data because of
"unnatural" precipitation or chemical regimes caused by the islands
themselves.

10.2.6. Summary and Recommendations

DMS seems to be the only sulfur component whose natural emission from
oceans is reasonably known. This gas may well be the most important
supplier of S to the marine environment, but this has not yet been
unequivocally established. Sulfur emissions from tidal flats also appear
to be rather well quantified. Otherwise, however, there is extremely
little information available on natural S emissions into the atmosphere.

The most critical data missing at the time of the Bermuda workshop
concerned:

1. The natural S emissions to the atmosphere from most land areas,
    most important of which are the H_2S concentrations and emis-
    sions in tropical continental ecosystems.

2. The production rates of SO_2 and H_2S by eruptive and noneruptive
    volcanoes; contributions from the latter source could be
    substantial.

3. The long-range transport of sulfur through the free tropo-
    sphere, especially the land-to-ocean transport of biogenic,
    volcanic, and anthropogenic sulfurs (H_2S, SO_2, $SO_4^=$).

4. The seasonal variabilities of S deposition in different eco-
    systems, especially for dry deposition to the oceans and wet
    and dry deposition to the continents.

We, therefore, recommend

1. That an expanded network of stations be established to measure
    emission and deposition on an ecosystem basis, especially over
    remote continental areas (cf. Chapter 9).

2. That observations be interpreted using appropriate meteorolog-
    ical analyses and, where appropriate, photochemical-model cal-
    culations.

3. That more research be concentrated on dry deposition and, in
    particular, on the measurement techniques used. The role of
    sea-salt and desert-dust particles as agents for dry deposition
    must also be investigated.

4. That laboratory studies of S-oxidation reactions in the gas
    phase be conducted in under realistic, i.e., low NO_x, environ-
    ments with particular attention focussed on reaction products
    and mechanisms of DMS + OH and DMS + NO_3 reactions.

5. That laboratory and field studies be continued on the oxidation
    mechanisms and rates of SO_2 in aqueous solutions.

10.3. NITROGEN CYCLE

10.3.1. Introduction

The discussion of the working group on nitrogen cycling focussed on two
themes: (1) our knowledge of surface-to-air exchanges of natural nitrogen compounds and (2) the origins of NH_4^+ and NO_3^- ions found in precipitation samples from remote regions.

10.3.2. Surface/Air Exchange of Natural Nitrogen Compounds

There were so few measurements of the surface/air exchange of gaseous and
particulate nitrogen compounds (i.e., NO_x, NH_3, NH_4^+, NO_3^-) that total
nitrogen emission could not be compared to deposition on a global or
regional basis. Known deposition values are significantly lower over
oceans than over land, indicating a significantly greater continental
source of reactive nitrogen than oceanic (Chapter 8). Again, based on
the available deposition measurements, the tropospheric nitrogen cycle is
more active in the tropics than in higher latitudes.

10.3.2.1. Nitrogen Oxides. Only one measurement of NO production in the
ocean surface waters has been reported (Zafiriou et al. 1980). Considering the extreme variabilities of the nutrient content and the biological
activity of ocean surface waters—two factors that probably influence
production—an extensive measurement program would be necessary to determine this source more quantitatively. More measurements are needed
before designing such a costly monitoring effort. Similarly, there are
insufficient measurements of the NO_x exchange over soils or vegetated
surfaces, which should be at least as variable as that over ocean surfaces although with much higher fluxes. Thus, it was extremely difficult
to make an order-of-magnitude estimate of surface emission rates. There
are no measurements from the tropics or from the continents of South
America or Africa.

10.3.2.2. Ammonia. The question of the air/surface exchange of ammonia
gas, as pointed out in the review of nitrogen emissions in Chapter 2, is
not a simple one since it often involves an exchange between aqueous-
phase, physically dissolved ammonia ($NH_3 \cdot H_2O$) and atmospheric ammonia
(NH_3). This may occur between air and aqueous media situated in soil, in
the interior of plants, in deliquesced aerosols, or in the ocean-surface
layer. A direct source of ammonia not involving equilibrium is the decay
of organic matter, which on continents is probably the main source of
ammonia. The ocean is potentially both a source and a sink of ammonia.

Assuming that the ammonia concentration in the atmospheric marine
boundary layer is in equilibrium with dissolved ammonia in marine aerosols and in the ocean surface layer, one could calculate atmospheric
ammonia concentrations from observed variables. Equilibrium with the
ocean is described by the following two equations: In the ocean-surface
layer,

$$(NH_4^+) + (H_2O) \longleftrightarrow (NH_3 \cdot H_2O) + (H^+) \text{ and} \qquad (10\text{-}1)$$

between the ocean-surface layer and the atmosphere,

$$(NH_3 \cdot H_2O) \longleftrightarrow (NH_3)g. \qquad (10\text{-}2)$$

The first equation is a well-known ionic equilibrium with a pH_a of approximately 10. The second is a symbolic representation of Henry's Law.
The Henry's Law constant is well known for low ionic strength solutions
but has not been measured for seawater. Nevertheless, it is assumed that
the influence of salts has a minor effect and, therefore, can be neglected. The concentration of atmospheric NH_3 above an ocean surface
depends on the total dissolved-ammonia ($NH_4^+ + NH_3 \cdot H_2O$) concentration,
the temperature, and the pH of ocean water.   .
    The dependence of atmospheric ammonia concentration on total
dissolved-NH_3 concentration and temperature for a pH of 8.2 is shown in
Figure 10-2. Georgii and Gravenhorst (1977) used this figure to illustrate the expected range of atmospheric NH_3 above the ocean and to point
out that, except for a few outliers, their measurements over the Atlantic Ocean are within this range (i.e., 0.1-0.6 ppbv). Ayers et al.
(1984) report similar measurements off Tasmania, which are consistent
with the assumption of equilibrium at the ocean surface.

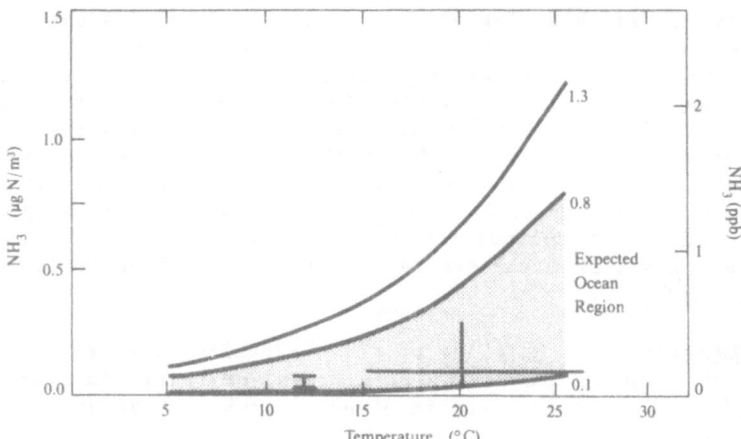

Figure 10-2. Dependence of Atmospheric Ammonia Concentration on Tempera-
    ture and Total-dissolved Ammonia in the Ocean-surface Layer. Calcula-
    tions assume equilibrium between air and ocean surface (shaded area)
    of pH 8.2. Data for Tasmania (at ~ 12° C) from Ayers et al. (1984);
    for the Atlantic (at ~ 20° C) from Georgii and Gravenhorst (1977).

The apparent mismatch between atmospheric NH_3 concentrations found at a particular location in the marine boundary layer and those calculated from aerosol composition measured at another location need to be verified by new data from comprehensive well-designed field experiments.

10.3.3. Origins of Ammonium and Nitrate in Precipitation in Remote Marine Areas

10.3.3.1. Ammonium. The NH_4^+ found in precipitation samples from remote marine areas, such as Amsterdam Island (Galloway and Gaudry 1984), probably originates from two sources: (1) from aerosols in the marine boundary layer and (2) from gaseous ammonia in the ocean-surface layer. Assuming that NH_3 gas concentrations are 50 ppt (v) and NH_4^+ aerosol concentrations are 0.1 $\mu g/m^3$ of air, a volumetric scavenging ratio of 5-10 $\cdot$ 10^5 (typical of that for highly soluble gases and aerosols) yields an NH_4^+ concentration in rain of approximately 4.1 $\mu eq/l$ to 8.2 $\mu eq/l$ (3/4 from the aerosol, 1/4 from the gas phase). Thus, observations at Amsterdam Island of 2 $\mu eq/l$ to 4 $\mu eq/l$ NH_4^+ in rain could be easily explained by the scavenging of gaseous and particulate ammonia.

10.3.3.2. Nitrate. Our discussions about the origin of precipitation NO_3^- in remote marine areas were still speculative because of the uncertainties in our understanding of natural NO_x sources. We felt that NO_3^- in rain originates from HNO_3 or aerosol NO_3^- derived from HNO_3 and that HNO_3 originates from nitrogen oxides from the following sources.

1. Continental surfaces: NO_x has to be transported into the free troposphere before conversion to HNO_3 or, otherwise, it would probably be rapidly depleted in the boundary layer by dry deposition and precipitation scavenging.

2. Lightning: either from cloud-to-cloud lightning in the upper troposphere or from cloud-to-ground lightning occurring a few days before any precipitation.

3. Destruction of N_2O: Estimates based on available N_2O measurements and photochemical theory indicate that the contribution from the destruction of N_2O in the stratosphere is an order of magnitude smaller than that from lightning.

4. Ocean surfaces: Although this source seemed very unlikely, more measurements are needed before discarding it as a possibility.

To answer the many questions about the sources of NO_x to remote marine areas, we suggest the following experiment: That the vertical profiles of nitrogen species HNO_3, PAN, NO_x, aerosol NO_3^- and total odd nitrogen be measured in the air flowing from remote continental regions (i.e., Brazil or the Chilean desert) over remote marine areas.

10.4. SUMMARY

Measurements of the air/surface exchange of NH_3 and NO_x are needed.
There is little information available for NO_x and virtually no information for NH_3 that can be used to assess global emissions.

The validity of the current Henry's Law theory for ocean-surface
water should be checked in a well-designed field experiment that would
include measuring aerosol composition.

The concentrations of NH_4^+ and NO_3^- found in rain from remote marine
areas are consistent with the reported concentrations of gaseous and
particulate matter and the current precipitation scavenging theory. The
origin of precipitation NO_3^- is not well known although lightning and
continental NO_x emissions are likely sources.

We recommend a field experiment in which the vertical profiles of
concentrations of NO_x, PAN, HNO_3, aerosol NO_3^-, and total nitrogen would
be measured in air flowing from a remote continental area with low NO_x
emissions over a remote marine area (i.e., from the Chilean desert west-
ward over the upwelling Pacific).

10.5. REFERENCES

Andreae, M. O. 1985. The ocean as a source of atmospheric sulfur com-
        pounds. In The Role of Air-Sea Exchange in Geochemical Cycling (P.
        Buat-Ménard, P. S. Liss, and L. Merlivat eds.) Dordrecht:Reidel (in
        press).
Andreae, M. O., and H. Raemdonck. 1983. Dimethylsulfide in the surface
        ocean and the marine atmosphere: A global view. Science 221:744-747.
Andreae, M. O., R. J. Ferek, F. Bermond, K. P. Byrd, R. T. Engstrom, S.
        Hardin, P. D. Houmere, F. LeMarreck, R. B. Chatfield. 1985. Di-
        methylsulfide in the marine atmosphere. J. Geophys. Res. (in press).
Ayers, G. P., J. L. Gras, A. Adriaansen and R. W. Gillet. 1984. Solubil-
        ity of ammonia in rainwater. Tellus 36b:85-91.
Bonsang, B., B. C. Nguyen, A. Gaudry, and G. Lambert. 1980. Sulfate
        enrichment in marine aerosols owing to biogenic gaseous sulfur
        compounds. J. Geophys. Res. 85:7410-7416.
Galloway, J. N. and A. Gaudry. 1984. The composition of precipitation on
        Amsterdam Island, Indian Ocean. Atmos. Environ. 18:2649-2656.
Galloway, J. N., G. E. Likens, W. C. Keene, and J. M. Miller. 1982. The
        composition of precipitation in remote areas of the world. J.
        Geophys. Res. 87:8771-8786.
Georgii, H. W., and G. Gravenhorst. 1977. The ocean as source or sink
        of reactive trace gases. Pageoph. 115:503-511.
Zafiriou, O.C., M. McFarland, and R. H. Bromund. 1980. Nitric oxide in
        seawater. Science 207:637-639.

11. SUMMARY

James N. Galloway
Department of Environmental Sciences
University of Virginia
Charlottesville, VA

Robert J. Charlson
Department of Civil Engineering
Environmental Engineering and Science Program
University of Washington
Seattle, WA 98195

Meinrat O. Andreae
Department of Oceanography
Florida State University
Tallahassee, FL 32306

Henning Rodhe
Arrhenius Laboratory
Department of Meteorology
University of Stockholm
S-10691 Stockholm, Sweden

11.1. INTRODUCTION

This book is the result of a five-day workshop concerning the atmospheric cycling of sulfur and nitrogen in the remote atmosphere. The workshop, attended by 24 scientists from 9 countries, was structured around the atmospheric cycle and consisted of working groups on emission, transformation, transport, and deposition. Each working group used a background paper prepared before the workshop by the chairman of the group. Following the workshop, the rapporteur of each working group prepared a document summarizing his group's views.

The objective of the workshop was to assess our level of knowledge on the atmospheric cycling of sulfur and nitrogen in remote areas and to make recommendations for research.

Workshop participants were selected for their expertise in the areas of sulfur and nitrogen cycling in the remote atmosphere. They were encouraged to bring not only summaries of their published work but also unpublished data sets. Thus, the workshop provided an opportunity for the rapid and efficient exchange of data and ideas and gave us an opportunity to synthesize sulfur and nitrogen cycling in the remote atmosphere. This chapter is a summary of that synthesis.

J. N. Galloway et al. (eds.), The Biogeochemical Cycling of Sulfur and Nitrogen in the Remote Atmosphere, 215–224.
© 1985 by D. Reidel Publishing Company.

11.2. EMISSIONS

11.2.1. Summary

- Fundamental difficulties, both conceptual and technological, stand in the way of obtaining accurate flux measurements from most important environments and ecosystems, e.g., oceans and forests.

- The only reasonably well-characterized flux of a sulfur species from a biological source is that of DMS from the oceans. The estimates of sulfur emissions from continental regions are the most uncertain because of both analytical problems and the inadequate coverage of subtropical and tropical regions.

- The biological and ecological determinants that control the emissions of most sulfur species are largely unknown.

- Volcanic emissions of sulfur gases occur predominantly during eruptive phases. This episodic character and the remote location of many volcanoes make a global-emission estimate rather uncertain. Satellite data should be used more frequently.

- A best-guess estimate for the global emission of natural sulfur is about 80 Tg S/yr (40 Tg from ocean, primarily as DMS, 30 Tg from the continental biosphere, and 12 Tg from volcanoes). The uncertainties of these estimates are as high as a factor of 2.

- Ammonia emissions from almost all important ecosystems are essentially unknown, primarily because of analytical and sampling difficulties.

- The biological/ecological parameters controlling the emission of nitrogen species are better understood than than those controlling the emission of sulfur species. Nevertheless, because of the lack of data coverage for many important regions and the problems with flux measurement techniques, the magnitude of the fluxes cannot be accurately estimated.

- Lightning-produced NO_x must be expected to influence substantial regions of the troposphere because of its production in the upper equatorial and tropical troposphere where removal rates are slow.

- The emissions of S and N compounds from biomass burning are expected to make significant contributions to the global cycle. They need to be further investigated in terms of both the species and the quantities emitted.

11.2.2. Recommendations

- Comprehensive protocols should be developed for both constituent analyses and flux measurements.

- Different flux-measurement techniques should be intercompared.

- Areal variabilities of emissions within, as well as those
  between, ecosystems should be determined.

- Emissions of all relevant S and N species should be measured in
  tropical ecosystems and tropical ocean regions.

- Specific additional emphasis should be placed on obtaining NH_3-
  emission measurements.

- Field studies of NO_x production in lightning should be
  conducted.

- Total emission rates of S and N should be established so that
  the presence or absence of emissions from presently unstudied
  species may be determined.

11.3. TRANSPORT

11.3.1. Summary

Past research has shown that a simple meteorological evaluation of
remote atmospheric chemistry data can, in many cases, be performed using
back-trajectory analyses. Whereas the uncertainty of estimated individual trajectories is very substantial, the statistics of many trajectories may give reasonable information about the origin of air masses,
several days back, affecting a particular site. The use of anthropogenic, as well as natural, tracers is another potentially powerful method
of identifying the nature and the location of sources of a measured
species. Detailed atmospheric models should only be used when sufficiently large meteorological and chemical data bases are available.

Some attempts have been made to quantify the long-range transport of
S and N species, for example, from industrial regions to certain remote
locations. In general, however, our lack of quantitative knowledge about
transformation and removal processes, as well as about transport processes, limits the possibility of making such estimates at the present
time.

A framework of meteorological regimes was developed during the
meeting. This categorization is tailored to the chemical cycles of the
atmosphere-earth system. It is hoped that this framework can be used by
scientists investigating emission and transformation questions to organize their study of S and N cycles.

During the group discussions, several simple meteorological sce-
narios were defined and recommended for further study. Because these
are situations with reasonably well-known meteorological parameters, the

emission, transformation, and deposition processes involved can be
studied without unnecessary complications.

11.3.2. Recommendations

We have recommended the following experiments to further our understanding
of the transport of sulfur and nitrogen species through the remote
atmosphere.

- The study of vertical transport in an equatorial, precipitating
  cumulonimbus cloud. This experiment would measure the vertical
  transport of surface air into the upper troposphere together
  with the natural emissions in the area using an artificial
  tracer. Measurements of wet and dry deposition should be
  coupled with this experiment.

- The study of the transition from a marine to a continental
  regime, or vice versa, in a region of a steady onshore or offshore flow. This experiment should be designed to look for any
  modification of the marine (continental) chemical characteristics of the atmosphere as a function of distance inland (offshore). It should first be performed during a dry period and
  then during a wet period. Possible areas of the world for the
  onshore study are the Olympic Peninsula in the northwestern
  United States and the Andes Mountains in Chile. Offshore
  possibilities would be eastward from Africa and eastward from
  North America. A small-scale version of this experiment, done
  under steady-state conditions (e.g., orographic precipitation),
  would be ideal for the study of cloud chemistry and wet deposition and could be located in the coastal mountains of Norway.

- The study of budgets and cycles in a limited region. This
  experiment should be explicitly designed to link the studies of
  emission, transformation, and deposition to that of transport.
  A well-defined geographical region with little expected spatial
  heterogeneity in any of these processes should be selected.
  Although logistically the North Atlantic would be the ideal
  location, the atmosphere cannot be considered to be remote from
  man's activities.

11.4. TRANSFORMATIONS

11.4.1. Summary

- We do not know the vertical or horizontal distributions of NH_3
  in remote or polluted areas. More measurements of NH_3 concentrations are urgently needed, particularly over remote continental areas. Removal processes must be better defined.

- Ozone, nitrogen oxides, and J values are relatively uniform and
  sampled with adequate frequency when large P values are observed. These P values suggest that there are unknown

reactions that may greatly affect our understanding of
reactive-N chemistry and ozone-production chemistry.

- The cause of the discrepancy between the sum of all the identifiable nitrogen compounds and the measurement of total nitrogen
  in an air sample that can be reduced to NO must be found before
  definitive answers concerning the transfer of nitrogen compounds in the atmosphere can be pursued.

- Does an additional reaction (possibly DMS + NO_3) occur that
  constitutes a sink for reactive N? Is this consistent with the
  measured-N budget? These questions need to be resolved.

- Are there any tropospherically significant reactions of reactive nitrogen in cloud or aerosol solution? This question also
  needs to be answered.

- Atmospheric H_2S should be measured from airplanes high above
  continental areas.

- There appear to be sufficient oxidants for S (IV) in remote
  cloud water in clean areas, provided inhibition does not occur.
  However, radical and redox reactions may contribute to more
  direct oxidation or peroxide production. They may also alter
  the oxidation in as yet unperceived ways.

- Major pathways for the physical and chemical transformations
  appeared to be understood qualitatively and agreed upon. However, important new pathways might be discovered.

- Simultaneous observations should be taken of all or most of the
  relevant species to determine their physical state. Tests must
  be devised to allow the observers to follow the atmospheric
  life of an atom of sulfur or nitrogen from its source to its
  sink. This would necessarily require an understanding of both
  the chemical and the physical transformation processes.

- Laboratory studies should be conducted to examine individual
  processes under a realistic range of chemical and physical
  conditions. Field observations should be compared to test
  hypotheses.

- A variety of compounds, e.g., dimethyl sulfoxide and organic
  nitrates, which are suspected, should be identified and
  quantified.

- The existence of unsuspected compounds should be explored by
  assaying the total sulfur and non-N_2 in nitrogen in air samples
  and all the compounds that contribute to the totals should be
  identified.

11.4.2. Recommendations

We agreed that the following hypotheses should be tested:

   1. **Liquid–phase oxidation apparently proceeds significantly in a clean troposphere.**

SO_2 oxidation does proceed significantly by OH radical oxidation in clean air. The main questions regarding SO_2 oxidation by remote tropospheric clouds probably concern the transport time of an SO_2 molecule to a cloud. This meteorological transport may be a limiting rate.

   2. **Long–range transport of nitrogen is by HNO_3 and PAN.**

HNO_3 and PAN have the longest lifetimes of the nitrogen species we considered and are, therefore, the candidates for the long-range transport of nitrogen to oceanic and other remote regions. Particulate nitrate, which is known to reside in the 1-6 μm range, will fall out by gravitational settling and cannot transport nitrogen over very great distances. Therefore, we concluded that particulate nitrate in remote oceanic regions is formed in situ from the HNO_3 precursor reacting with the basic sea-salt aerosol.

   3. **Upper tropospheric tropical lightning is a source of NO_x.**

Upper tropospheric tropical lightning may be a major source of the background NO_2 observed in remote oceanic areas based on an estimated yield of NO_x in typical cloud-to-ground flashes and satellite observations of lightning-flash frequency. The amount of nitrogen annually produced above 10 km in the tropical zone might be sufficient to account for the observed deposition to remote ocean regions.

   We also felt that the following conclusions need further investigation:

   • Resistances to transport in air and at the (cloud or aerosol) droplet surface and within the droplet must be considered. Laboratory reaction rates that are fast (time scales of minutes) must be checked. The equations of transport and reaction must be solved simultaneously outside and within the droplet.

   • Clearly, pH is very sensitive to both NH_3(total) and $SO_4^=$ aerosol under background conditions of approximately 0.1 μg/m^3 NH_3(total) and a fraction of a μg/m^3 $SO_4^=$. Thus, if any reaction rate in cloud water is pH dependent, the levels of both NH_4^+ and $SO_4^=$ must be expected to have control.

   • More measurements of hydrogen peroxide are required, especially in the tops of clouds where significant quantities of H_2O_2 can be formed by radical and ionic reactions in values HO_2.

However, radical and redox reactions may contribute to more
direct oxidation or more peroxide production. They may also
alter the oxidation in as yet unperceived ways. Because these
physical processes are consecutive with the initial chemical
reactions of key S and N species and because they are competitive with dry-removal processes to the surface, the question
remains as to the overall lifetime prescribed atmospheric
conditions.

11.5. DEPOSITION

11.5.1. Summary

- In remote marine, continental, and polar areas, the wet deposition of excess S is usually more than twice that of the dry
  deposition. This partition, however, depends on the rainfall
  rate in each region. In some remote continental areas, like
  deserts, dry deposition may be the most important removal
  mechanism.

- In remote marine areas, the wet deposition of sea-salt S appears
  to be several times greater than that of the dry deposition.
  The dry-deposition rates of SO_2 and excess $SO_4^=$ are of similar
  magnitude.

- In remote continental areas, the dry-deposition rates of SO_2
  could be twice that of $SO_4^=$.

- In polar areas, the wet-deposition rate of excess $SO_4^=$ is
  greater than the dry-deposition rate.

- In remote marine areas, the wet-deposition rates for NH_4^+ and
  NO_3^- are approximately equal to and several times greater than
  dry-deposition rates for either the oxidized-N compounds or the
  reduced-N compounds. Although the estimates of the rates of
  dry deposition for all N species vary widely, it is unlikely
  that dry deposition will become as important a sink for N
  species as wet deposition.

- In remote continental areas, the wet deposition of NO_3^- is
  about equal to the sum of the dry deposition of NO_3^-, HNO_3, and
  NO_x. For NH_4^+, however, the wet deposition is several times
  greater than the dry deposition of NH_3 plus NH_4^+ aerosol. For
  the N species as a whole, the estimated wet-deposition rate is
  about double that of the dry-deposition rates of N species
  considered here, whether the species is oxidized or reduced.

- In polar areas, the rates of wet deposition of NO_3^- and NH_4^+
  are approximately equal. There are no estimates of dry-
  deposition rates available.

- The deposition of NO_3^- and NH_4^+ to polar areas is about an order of magnitude less than the deposition to remote marine and continental areas. The deposition of NO_3^- and NH_4^+ to remote marine areas is essentially equal to the deposition to remote continental areas, given the errors involved in ascertaining deposition rates.

11.5.2. Recommendations

11.5.2.1. Temporal and Spatial Representativeness.

At this time, the data base on atmospheric deposition is inadequate to serve as the basis for designing a global deposition network. As an initial minimal configuration, such a network should consist of at least one station in each major wind regime in each continental and ocean region. The data base could be broadened dramatically if some existing networks, such as BAP-MoN, were to adhere to a suitable deposition-sampling protocol.

There is a critical need for long-term measurements in the Southern Hemisphere. Such measurements would serve to establish background values and, over the longer term, to determine trends that could be related to anthropogenic impacts. In some ocean regions, there are no islands suitably located to serve as sampling stations for some major wind regimes. In such cases, efforts should be made to establish sampling programs on ships of opportunity.

11.5.2.2. Integration of Wet-deposition Data with Other Atmospheric Cycling Measurements.

The deposition process is but one aspect of the entire atmospheric cycle of S and N. Deposition data become more meaningful if they are accompanied by data related to the other processes--emissions, transport, and transformation. Deposition measurement programs should be planned accordingly. Important deposition-related aspects of these processes are discussed in other sections.

At this point, we want simply to emphasize the need for deposition measurements in areas that are representative of major source types and source regions whether natural or anthropogenic. It is especially important that deposition measurements be made in conjunction with emission measurements.

11.5.2.3. Measurements Critical to Reducing Uncertainties.

- Precipitation samples should be collected on an event basis using rain collectors that exclude dry deposition. Precautions should be taken to maintain the chemical integrity of the samples.

- As a start, long-term collection sites for wet deposition and atmospheric gases and aerosols should be established in each of the 10 major environmental zones.

- New types of shipboard collectors must be designed and tested.

- Organic N and organic acids should be measured in precipitation from remote areas.

- The regional representative collection sites should be evaluated in those areas where rainfall gradients are large or where localized sources are strong.

11.5.2.4. <u>Integrate Measurements of Atmospheric Constituents with Wet Deposition Measurements</u>. The concentration of many aerosol- and gas-phase species should be measured in conjunction with wet-deposition studies. For some species, dry-deposition velocities can be computed on the basis of such data. Aerosol data would also be useful in transport studies. These data are especially important because many meteorological events have a low-precipitation probability. It may also be desirable to sample for selected gas species, such as SO_2, using chemically treated filters.

11.5.2.5. <u>Isotopes</u>. Efforts should be made to develop source-identification techniques based on the isotopic ratios of $^{34}S/^{32}S$ and $^{15}N/^{14}N$. Isotope ratios are especially good tracers because the chemical behavior of isotopes is identical. Fractionation occurs because of differences in physical properties from the different isotopic masses. The isotopic ratios in a specific sample reflect those of the source material modified by possible fractionation processes. For example, some major sources for atmospheric sulfur compounds (e.g., sea salt, volcanoes, biological H_2S) have a characteristic isotope ratio. Unfortunately, some other sources (e.g. coal combustion) have less specific isotope ratios. Because of the small elemental mass difference between ^{34}S and ^{32}S and between ^{15}N and ^{14}N, the isotope ratios are not changed appreciably by reactions occurring during atmospheric transport.

Very little work has been done on the development of N isotope techniques. Such work should be encouraged. Experiments should be carried out to characterize the source areas. The isotopic composition of deposition samples from remote regions should be determined so that any possible trends can be related to sources. Some values for the marine precipitation and aerosol samples and for stratospheric aerosols have been published. However, extensive surveys of both source areas and sink regions are needed.

11.5.2.6. <u>Ecosystem Approach</u>. The principal advantage of an ecosystem approach is that the value integrates the complexities of the various real components for a natural area. This approach has several disadvantages: It is labor intensive and expensive and the accumulated errors may be large—the computed net input is small and is the difference between two large values.

It is recommended that, where possible, measurements of wet deposition be made in conjunction with elemental budget studies of ecosystems to allow not only the determination of removal from one reservoir but also to establish the significance of the input to the receiving reservoir.

11.5.2.7. <u>Other Important Chemical Constituents</u>. Deposition measurements should be extended to those species important to our understanding of the sulfur and nitrogen cycles.

The existing wet-deposition data sets consist almost exclusively of measurements of the principal oxidative end members of the S and N cycles: $SO_4^=$ and NO_3^-. However, measurements of other important species are needed to understand the entire cycles of these elements. Generally, these measurements pose a greater challenge to our analytical capabilities. All of the following constituents should be measured in air, clouds, and rain.

- Dimethyl sulfoxide: This is a suggested oxidation product of $(CH_3)_2S$.

- Methanesulfonic acid: This measurement is very important since CH_3SO_2OH is a significant product in the oxidation of dimethylsulfide, at least under laboratory conditions.

- Hydroxy methanesulfonic acid: This species is produced as an aqueous complex of formaldehyde and SO_2. Its presence would indicate a substantial inhibition of SO_2 oxidation.

- Formaldehyde: This component is not only important because it acts as an inhibitor of SO_2 oxidation but also because it significantly affects the OH radical concentrations and provides a check on the carbon cycle.

- Hydrogen peroxide: This compound, along with O_3, is thought to be the major oxidizer of SO_2 in solution. It also plays a major role in determining HO concentrations. Organic peroxides could be almost as important and essentially nothing is known about them.

- Ammonia and the amines: These interact strongly with the sulfur cycle in clouds and aerosols.

- Organic nitrogens: These are suspected to be major carriers of N in the remote atmosphere. Their concentration is unknown but could be comparable to the concentration of all other species of reactive and reservoir nitrogen.

- Acetone: This is an indicator of the recent influence of nonmethane hydrocarbons on air chemistry.

- Formic and acetic acids: These are major sources of free acidity in precipitation from remote regions and thereby control, in part, all pH-dependent transformations of S and N species.

EPILOGUE

What have we learned?
What should happen next?
How should we accomplish these future efforts?
Was it worth the effort?

On return flights with participating colleagues, in telephone conversations pursuant to writing this volume and at various meetings in far-flung places over the several months since Bermuda, these four questions arose. Although the whole group has not met since October 1984, comments at the meeting and in this later communication provided the organizing committee with solid answers.

First, it is clear that the notion of a biogeochemical cycle as an organizing framework provided a functional basis for productive discussions. The atmospheric parts of the cycles of sulfur and nitrogen really are not themselves complete biogeochemical cycles. But, because (with few exceptions) sulfur and nitrogen compounds do not accumulate in the atmosphere, we can safely assume that there is a mass balance of inputs and outputs. Sources and sinks can be compared, and magnitudes of fluxes can be estimated, for example, in the complex transformations between species and between phases in the atmosphere. Indeed, it is the integration of the entire picture of these parts of the sulfur and nitrogen cycles and of their interactions that was the overall purpose of the meeting. Thus, sources, transformations, transport, and removal are linked together within this fundamental cycle concept. That successful communication occurred, in spite of the diverse backgrounds of participants, is attributable to the effectiveness of the cycle approach.

Second, we learned that there is an increasing degree of internal consistency in the data bases describing the atmospheric cycles of these two key elements. There are gaps and a few mismatches, but clearly, progress has been made over the past decade, e.g., since the publication of SCOPE 7. However, viewing these cycles in approximate steady-state is only a beginning. What will be needed next is the development of a sound understanding of the factors controlling the cycles, their sensitivities to changes, and the possibilities of secular trends in concentrations and fluxes. Only when a sound understanding of the system dynamics has been established will it be possible to forecast the effects of regional- or global-scale perturbations, e.g., from human activities or natural occurrences.

Third, it is clear that much more understanding can be gained by simultaneous observations of different parts of these cycles than is possible when the different pieces are studied in isolation. For example, the observation of the removal flux of a particular species can never lead to a complete picture of the overall behavior of that species in the atmosphere or of its importance to the removal flux itself. Thus, just as the meeting produced a useful integration and intercomparison of data, future experiments should be designed to provide the possibility of integrating the data into as complete a picture as is practical and affordable. It is also clear that multiple observations of this sort cannot be accomplished by solo investigators but rather that teams of cooperating investigators are mandatory if the needed variety of measurement methods are to result. Obviously, these cycles encompass processes that occur over the entire globe and international cooperation will be required for successful studies to take place.

Thus, there is a compelling need for cooperative, international programs aimed at providing the necessary scientific expertise and geographical coverage. Although it was not an aim of the Bermuda meeting to organize such programs, the high level of effective scientific communication at the meeting demonstrated that the scientists themselves might very well be able to provide the needed coordination for future studies. There should be ample opportunities for bringing together the appropriate scientists with adequate frequency. Several international meetings within the coming years will have such cooperative research on the agenda.

It is our hope that the high level of cooperation and communication that was evident at the Bermuda meeting will continue to develop as cooperative programs and projects become a reality.

Finally we might ask, Why bother to go to such efforts? Although the Bermuda meeting did not consider practical and applied aspects, this multidisciplinary science of ours is central to a wide range of socially important issues. To name a few,

- acid rain,
- ozone and other photochemical oxidants,
- air–pollution transport between countries,
- nutrient delivery from the atmosphere to the biosphere.

We do not want to claim at this point that we will be able to solve all the scientific problems related to these issues. What we can state unequivocally, however, is that, if we do not develop the necessary scientific understanding, we will not be able to provide a scientific basis for solving the problems at some time in the future.

Finally, it may be useful to view the development of this integrative science of ours in contrast to the usual ways that science has developed over the last century or so. In the context of a single–discipline science, success often can be achieved by figuring out how to divide a problem into smaller and smaller parts, each of which provides "well–posed problems" for research. More and more is learned about some phenomenon with less and less general applicability. It is evident that the study of biogeochemical cycles employs both this traditional reductionist approach and a requirement for fitting the result back into the larger, integrated picture.

Robert J. Charlson
James N. Galloway
Meinrat O. Andreae
Henning Rodhe

June 1985

AUTHOR INDEX

Reference lists at the end of each chapter are not included in this index.

SUBJECT INDEX

See also references to individual sulfur and nitrogen compounds have
been omitted to conserve space. Most sulfur and nitrogen compounds
referred to in the text are individually indexed.